Standard

AIRCRAFT HANDBOOK

Standard AIRCRAFT HANDBOOK

Revised By

LARRY REITHMAIER

Third Edition

1980

Originally Compiled and Edited by

STUART LEAVELL

STANLEY BUNGAY

AERO PUBLISHERS, INC.

329 W. Aviation Road Fallbrook, CA 92028

Library of Congress
Catalog Card No. 80-67736

ISBN 0-8168-8502-8

Printed and Published in the United States by Aero Publishers, Inc.

ACKNOWLEDGMENTS

We wish to thank the following organizations for their help in the preparation of this handbook: Adel Precision Products Corp.; Aircraft Tool Co.; Aluminum Co. of America; Beech Aircraft Corp.; Bell Aircraft Corp.; Boeing Aircraft Co.; Boots Aircraft Nut Corp.; Cannon Electric Development Co.; Cessna Aircraft Co.; Chance Vought Aircraft; Cherry Rivet Co.; Chicago Pneumatic Tool Co.; CONVAIR; Curtis Wright Corp.; Dill Manufacturing Co.; Douglas Aircraft Co.; Dzus Fastener Co.; Elastic Stop Nut Corp.; Glenn L. Martin Aircraft; B. F. Goodrich Co.; Grumman Aircraft; Huck Manufacturing Co.; Hi-Shear Rivet Co.; Hughes Aircraft Co.; Illinois Tool Works; Kaiser Aluminum; Lockheed Aircraft Corp.; Lufkin Rule Co.; McDonnell Aircraft Corp.; North American Aviation; Northrop Aircraft; Parker Appliance Co.; Piper Aircraft Corp.; Reed Roller Bit Co., Cleco Division; Republic Aviation Corp.; Reynolds Aluminum; Ryan Aeronautical Co.; Society of Automotive Engineers; Tinnerman Products, Inc.; and Zephyr Manufacturing Co.

The assistance of the following individuals in compiling, illustrating and proofreading has been a substantial aid and is gratefully acknowledged: Milton Brown, Vic Davis Julian Ehresman, 'Bob' King, Jack Viets, Gerald Whittemore and Linden Whittemore.

Stuart Leavell,
Stanley Bungay.

PREFACE TO THE THIRD EDITION

The Standard Aircraft Handbook was developed and first printed in 1952 as a general guide for aircraft workers, A&P mechanics and students. During the ensuing years, over one-quarter million copies of the first and second editions have been printed to attest to its usefulness and popularity.

Advancements in aeronautical technology dictate that a handbook used for both operational and instructional purposes, be under continuous review and brought up to date periodically. This third edition of the Standard Aircraft Handbook includes some revision to all chapters, however, the sections on riveting, bolts and fasteners, aircraft plumbing, aircraft electrical systems and materials and fabrication have been revised most extensively. In addition, the chapter on standard parts has been revised to include the latest standard specifications.

The subject matter of this handbook is considered sufficiently general to be applicable to any aircraft; however, this handbook is not intended to replace, substitute for, or supersede Federal Aviation Regulations or the manufacturers' instructions.

Additional data was provided by the Lockheed-California Company; Cherry Fasteners (Townsend-Textron); Dzus Fasteners Co., Inc. and the L.S. Starrett Co. Grateful acknowledgment is extended to these manufacturers for their cooperation in making material available for inclusion in this handbook. Also used was material from various FAA publications.

CONTENTS

Factory mechanics applying fuselage skin to a Learjet. *(Gates Learjet Corp. Photo.)*

CHAPTER I
RIVETING

METHODS OF RIVETING

One of the most important phases of the airplane assembly is the proper riveting of the various parts into a tight and efficient joint. The driving of the rivets is accomplished by any one of several methods, and the use of various tools to upset the rivet.

Pneumatic squeezers and pneumatic riveting hammers are the most common riveting tools an individual may use when working on the structure proper. The pneumatic squeezer is illustrated here in Fig. 1. This is a portable type and is used primarily to join stiffeners or brackets near the edge of a structure or around an opening, since its use is limited by the length of the jaws. The squeezer applies its force to the rivet by air pressure, acting on the large cylinder, which forces the jaws together with sufficient pressure to upset the rivet shank. One of the jaws is equipped with the proper sets to fit the rivet head while the other is flat to form the upset end.

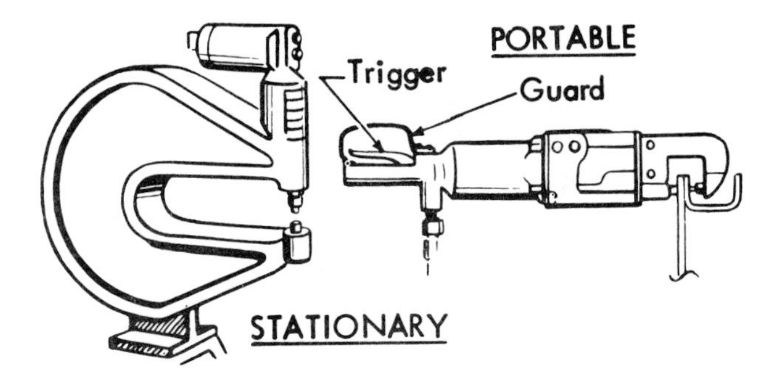

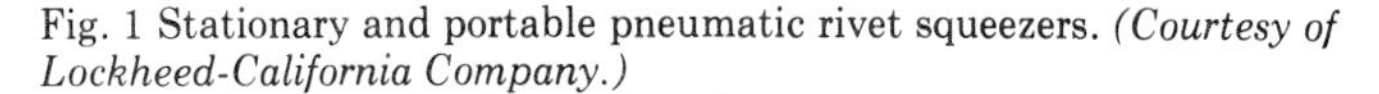
Fig. 1 Stationary and portable pneumatic rivet squeezers. *(Courtesy of Lockheed-California Company.)*

The pneumatic riveting hammer is illustrated in Fig. 2. It has the greatest use of all riveting methods due to its flexibility. Its driving action is obtained by air pressure driving a piston repeatedly against the rivet set which in turn applies its force to the head of the rivet. The rivet is upset by a bucking bar held solidly against the shank end of the rivet while the pneumatic hammer is operated.

RIVETING PRACTICE

Riveting with the pneumatic hammer generally requires partners; one driving the rivet and the other using a bucking bar which upsets the shank end.

The extensive use of this method of riveting requires skilled riveters in order to develop speed and efficiency in driving the many rivets used to join various structural parts and surfaces.

Of equal importance to the speed of riveting is, of course, to do this riveting in a manner which prevents damage to the airplane itself. This requires the constant attention and personal desire of the riveters to do good work.

The damage that may be done with the riveting hammer can readily be understood when you consider that the power of the riveting hammer is sufficient to upset the end of a rivet and could exert this same pressure against the surface of the airplane unless the driving action is confined to the rivet alone. Such damages occur when the rivet set slips from the head of the rivet or when the bucking bar is not held solidly in place while the hammer is being operated. Damage of this sort, even though only occasional can cause uncountable hours of rework or replacement of parts if not avoided.

Primary instructions to partners are illustrated in the following figures, which show a condition where the riveter and bucker may not communicate directly and must rely on signals to each other.

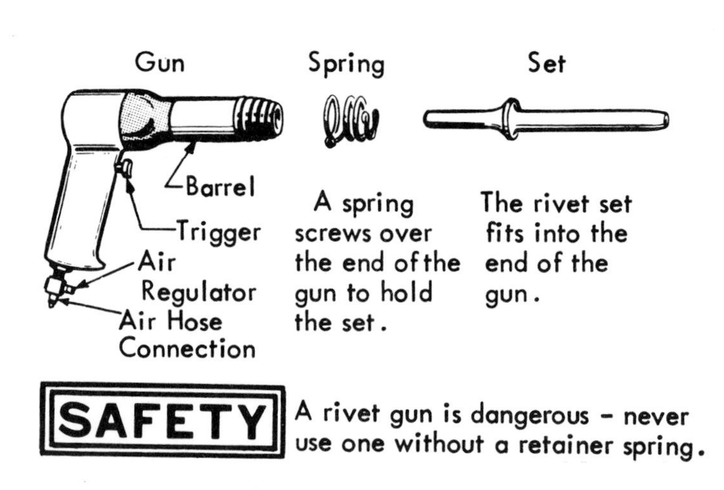

Fig. 2.—Pneumatic rivet hammer.

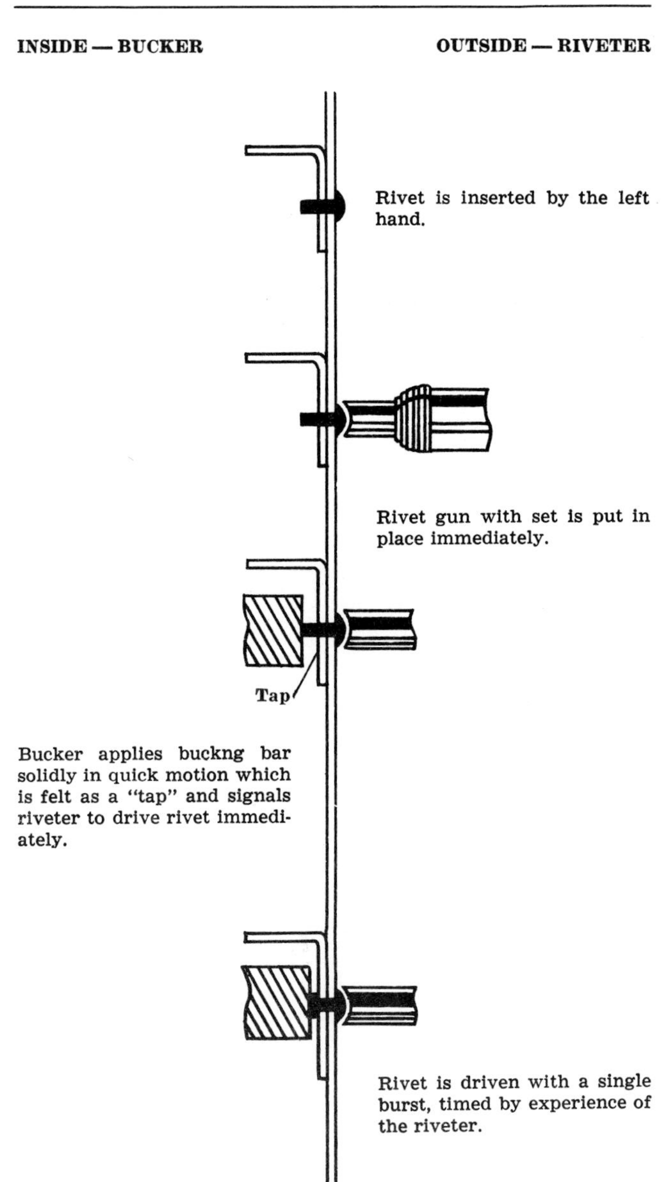

Rivet is inserted by the left hand.

Rivet gun with set is put in place immediately.

Bucker applies buckng bar solidly in quick motion which is felt as a "tap" and signals riveter to drive rivet immediately.

Rivet is driven with a single burst, timed by experience of the riveter.

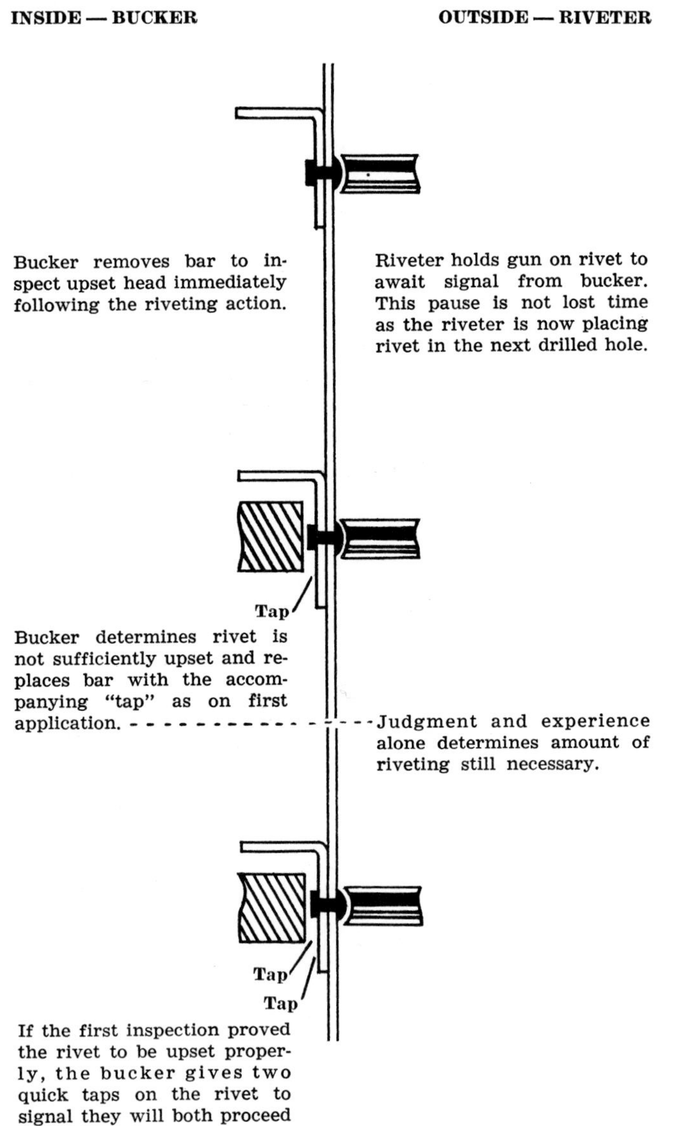
INSIDE — BUCKER
OUTSIDE — RIVETER
Bucker removes bar to inspect upset head immediately following the riveting action.
Riveter holds gun on rivet to await signal from bucker. This pause is not lost time as the riveter is now placing rivet in the next drilled hole.
Tap
Bucker determines rivet is not sufficiently upset and replaces bar with the accompanying "tap" as on first application.
Judgment and experience alone determines amount of riveting still necessary.
Tap
Tap
If the first inspection proved the rivet to be upset properly, the bucker gives two quick taps on the rivet to signal they will both proceed to the next rivet previously installed.

Although there are several steps necessary to upset each rivet, they are done in such rapid succession by experienced riveters that they may be accomplished as fast as the riveter can insert the rivets with one hand while driving with the other. It is easy to see possible mistakes, causing damage, during this speed of riveting. The following illustrations show common errors.

INSIDE — BUCKER **OUTSIDE — RIVETER**

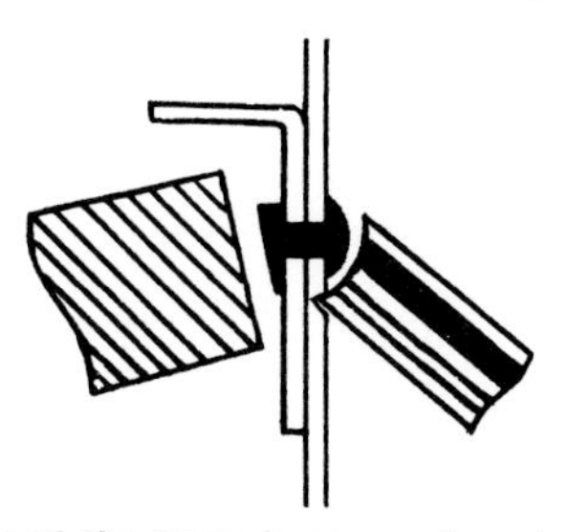

Both members of the team have erred — rivet must be replaced because of improper upset. The riveter has damaged the surface of the airplane with the edge of rivet set. Rework necessary here may necessitate special permission to use the part as is, or completely remove this section of the surface.

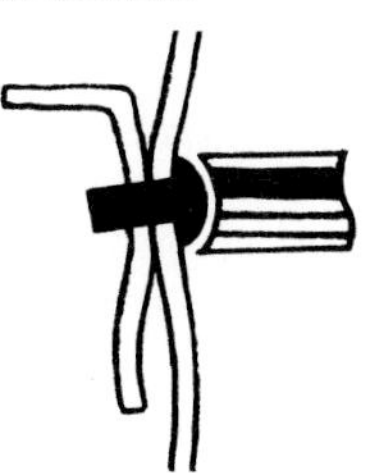

Any misunderstood signals may cause riveter to drive against rivet without the bar in place, thus dimpling the surface and bending sub-structure. Damage may be as great as above.

Rivet timing wrong — remove rivet.

Proper Driven Rivet.—Regardless of the upsetting or driving method, the resultant article should be consistently uniform. Figure 3A shows several specimens of driven rivets. Figure 3B shows standards for judging a good rivet.

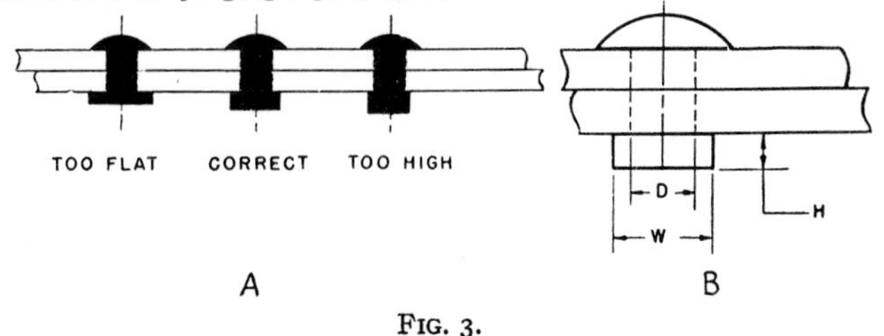

FIG. 3.

The width W should equal $1\frac{1}{2}$ times the original diameter D, and the height H should equal one-half the original diameter.

Correct Rivet Length.—To calculate the correct rivet length for a certain job, to the grip length or thickness of material through which the rivet must pass, add as follows:

Rivet dia., in.	Grip, in.	Add
$\frac{1}{4}$ or less	$\frac{1}{2}$ or less	$1\frac{1}{2}$ diameter of rivet
$\frac{1}{4}$ or less	Over $\frac{1}{2}$	$1\frac{1}{2}$ diameter + $\frac{1}{16}$ in. for every $\frac{1}{2}$ in. of grip
$\frac{5}{16}$ or more	1 or less	$1\frac{1}{2}$ diameter of rivet
$\frac{5}{16}$ or more	Over 1	$1\frac{1}{2}$ diameter + $\frac{1}{16}$ in. for every 1 in. of grip

The total is the proper length of rivet to use.

Remove Bad Rivets.—Occasionally a rivet is ruined and has to be replaced. To remove it, a hole is drilled through its head, with the same size drill used for the rivet hole, just deep enough to sever the rivet head from the shank (Fig. 4 A). The head is snapped off, then the shank is tapped from the hole with a pin punch, as shown in Fig. 4 B. If the head is carefully drilled, the shank may be pushed out with the drill.

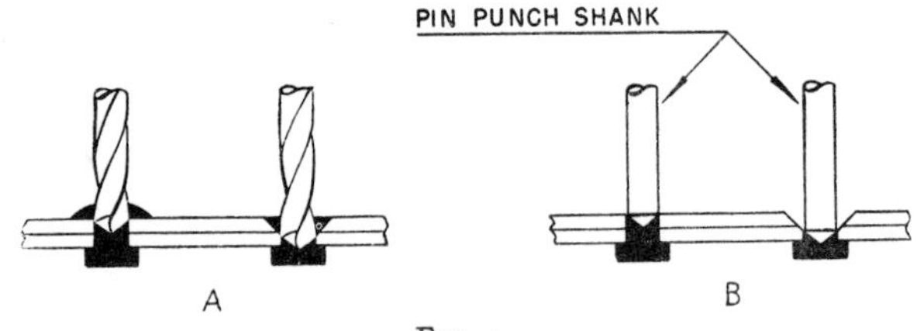

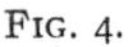

FIG. 4.

Drill Size for Rivets.—The most used rivet sizes with the recommended drill size for each are as follows:

Rivet Dia., In.	Drill Size	Rivet Dia., In.	Drill Size
$\frac{1}{16}$	No. 51	$\frac{5}{32}$	No. 21
$\frac{3}{32}$	No. 40	$\frac{3}{16}$	No. 10 or 11
$\frac{1}{8}$	No. 30	$\frac{1}{4}$	$\frac{1}{4}$-in.

RIVET TYPES AND IDENTIFICATION

Rivets differ and are identified by: (1) style of head, (2) material (aluminum alloy principally), and (3) size (diameter and length). To obtain a specific rivet, these three identifying terms must be called out by their proper names or by their proper identifying numbers as used on drawings. Rivets are identified by their MS (Military Standard) number which superseded the old AN (Army-Navy) number. Both designations are still in use however.

STYLE OF HEAD AND IDENTIFYING NUMBER

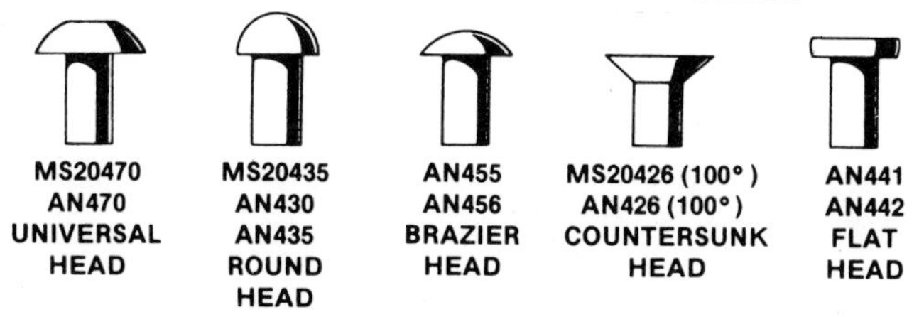

NOTE: When replacement is necessary for protruding head rivets—roundhead, flathead, or brazier head—they can usually be replaced by universal head rivets.

MATERIAL AND IDENTIFICATION
CODE BREAKDOWN

YOU CAN TELL THE MATERIAL BY THE HEAD MARKING

Rivet	Material Code	Head Marking	Material
	A	PLAIN (Dyed Red)	1100
	AD	DIMPLED	2117
	DD	TWO RAISED DASHES	2024
	B	RAISED CROSS (Dyed Brown)	5056
	M	TWO DOTS	Monel

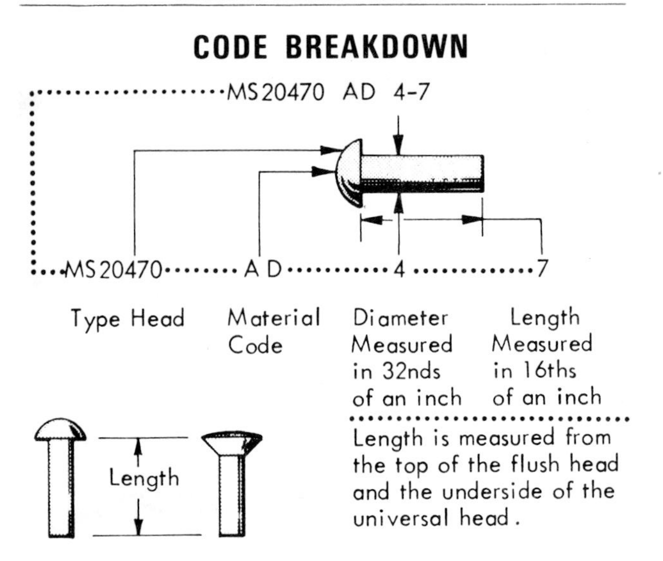

NOTE: The 2117-T rivet, known as the field rivet is used more than any other for riveting aluminum alloy structures. The field rivet is in wide demand because it is ready for use as received and needs no further heat-treating or annealing. It also has a high resistance to corrosion.

HEAT TREATED RIVETS

From the foregoing rivet identification, you will note the several aluminum alloys used for aircraft rivets. Some of these require special heat-treat processes up to and during the time of their use, and others are ready for use as received from the manufacturer. Therefore, it is important to be able to recognize the alloy from the MS or AN code and from the markings on the head of the rivet itself.

The rivets, ready for use, are from alloys 2117, 1100 and 5056. The alloy 2117 is actually a heat-treatable alloy but is fully heat-treated before use and requires no special attention by the riveter.

The rivets from alloys 2017 and 2024, if fully heat-treated, will be too hard and will crack upon driving unless used in a manner to complete the heat-treat process after driving. In order to do this it is desirable to understand a little of the heat-treat process. Read about the heat-treatment of aluminum alloys and the heat-treatment of rivets in Chapter V where it will be noted that these rivets must be stored in refrigerators until the time of their actual use. This cold storage prevents the completion of their heat-treatment, which will proceed when the rivets are removed from cold storage and their temperature rises to room temperature. This is an aging process which hardens the alloy after a certain amount of time at room temperature. This is automatic and the riveter must use the rivet within approximately thirty minutes after removal from the refrigerator. Be aware of the process, and understand that after this time the rivets are too hard for use. Rivets that have become hard before use will not be restored to their soft condition by replacing in the refrigerator, these must be returned for full heat-treat re-processing.

FLUSH RIVETING

Preparation of a surface for flush rivets is done by one of several methods: . . by countersinking, by dimpling, or by a combination of dimpling and countersinking. Regardless of the method of preparing the surface, the requirements are that the rivet be let into the surface until the rivet is either flush with, or slightly above the surface.

The rivet set used for flush riveting has a large, slightly contoured and highly polished surface to strike the rivet head. See figure 5.

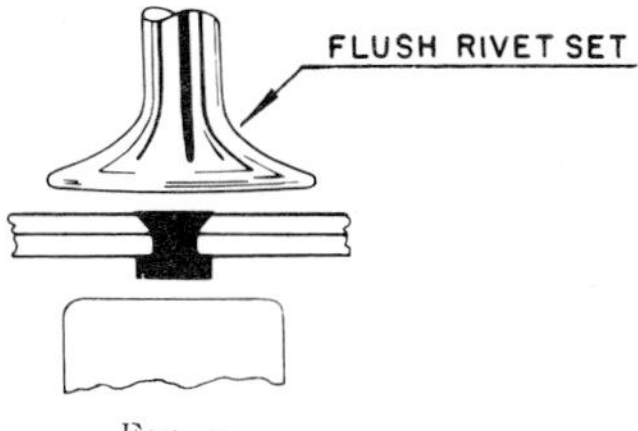

FIG. 5.

Countersinking for flush rivets is the most direct method of preparing the surface. The hole is first drilled to suit the diameter of the rivet shank, then beveled with the countersinking tool to suit the above mentioned requirements and the rivet is ready to be driven. (See the section on tools for the description of the countersinking tools).

The above Fig. No. 5 is a riveted joint completed in a countersunk surface.

Dimpled surfaces are prepared by the use of forming dies which actually dimple the material inward around the drilled hole until the flush rivet may be let into the surface and fulfill the requirements of flush riveting as above.

This formed section of the outer surface has a corresponding reverse dimple on the under side of the surface which is in turn nested into either a "dimpled" or a "countersunk" sub-surface. Fig. No. 6 below shows riveted joints in dimpled surfaces.

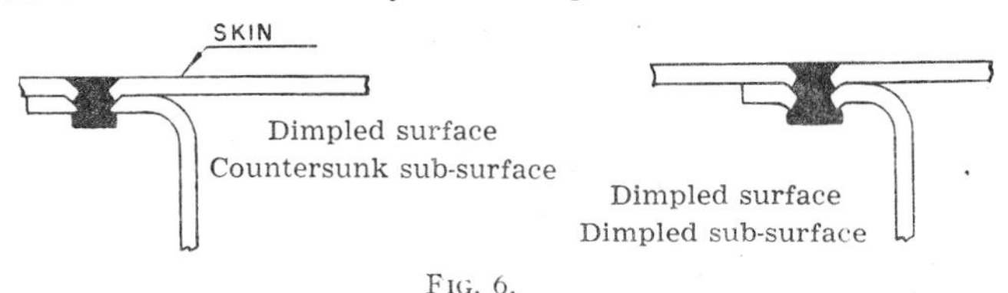

FIG. 6.

The dimpling dies for the above dimpling are used with both the squeezers and the pneumatic riveting gun and are designed especially for surface and sub-surface work. (Fig. 7).

The dimpling procedures will vary also according to individual practice; however, in general the surfaces are first fitted together and drilled to fit the rivet shank (or pre-drilled to a smaller size and reamed to full size after dimpling). They must then be dis-assembled for the dimpling operation on the outer surface and the dimpling or countersinking of the sub-surface.

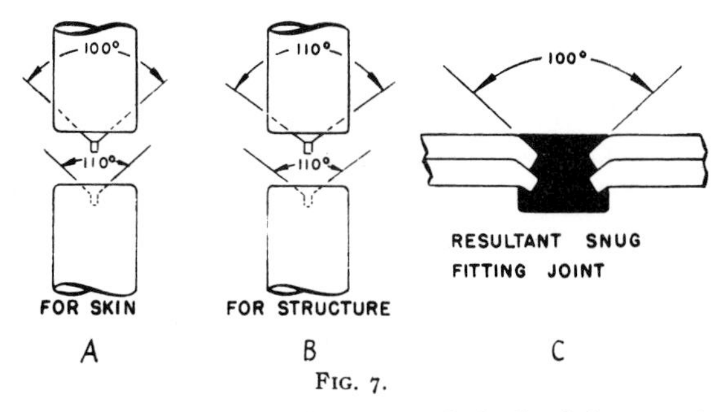

Fig. 7.

On some applications the sub-surface only is dimpled or countersunk when dis-assembled and then the outer surface is replaced and dimpled into the sub-surface. The sub-surface must be backed up by a draw bar and a bucking bar while the surface is dimpled with the form tool. (Fig. 8)

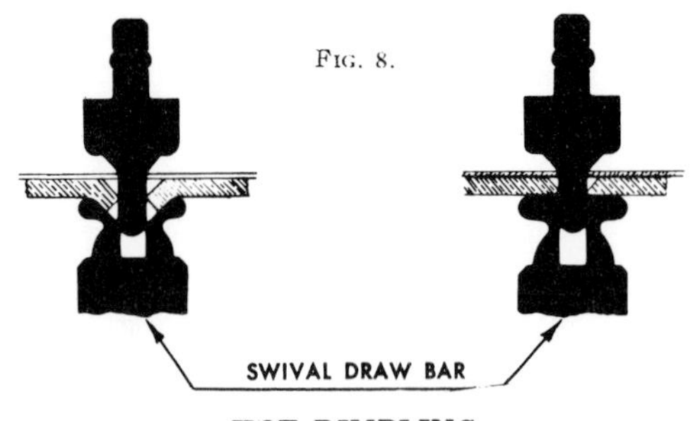

Fig. 8.

HOT DIMPLING

The above dimpling of surfaces and structures has limitations when used on the thicker materials and on the high strength aluminum alloy 7075.

The metal forming is quite severe in order to make the comparatively sharp bends required to make a flush surface with the rivet head.

Research and experimenting has developed "hot-dimpling" to enable dimpling of the stronger alloys without damage to the material because of the extreme forming.

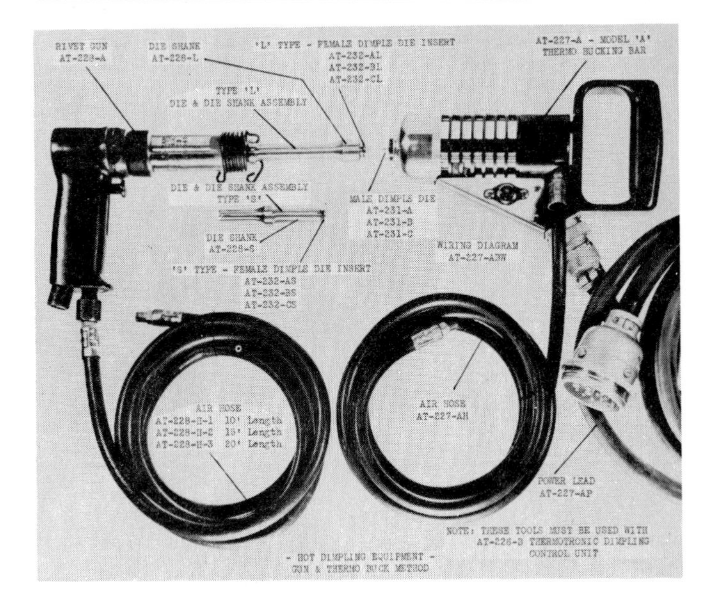

Fig. 9.—Hot dimpling equipment.

Hot dimpling is done by electrically heating the forming die which is held against the surface until the immediate area of the material is heated and the forming may be accomplished without damage to the material.

Naturally guess work must be eliminated from this type of work since too great heat will actually anneal the material to an extent which will destroy the strength characteristics of the material. This heating is electronically controlled to provide a dwell time to bring the material to proper temperature before the pneumatic hammer completes the forming operation. Trained men must be educated to the procedure and applications of the various airframe manufacturers.

RIVET MILLING (SHAVING)

A driven flush rivet, as stated above, will be flush with or slightly above the surface. The amount the rivet may be above the surface depends on the various manufacturers' practice; up to .015 of an inch is allowed for some use. However, on the wing leading edge and the front section of the fuselage, as well as on all surfaces of modern, supersonic aircraft, a near perfect flush surface is desired. This is accomplished by driving the flush rivet to suit the above requirements and then milling that portion of the rivet head which protrudes above the surface. Rivet milling is a hand operation as shown in the picture below and is done with a high speed motor driving a milling cutter. The tool is placed directly over the rivet and is plunged straight downward to cut the excess rivet head off. Adjustable stops are provided on the tool which are set to prevent the cutter from going deeper than the surface proper.

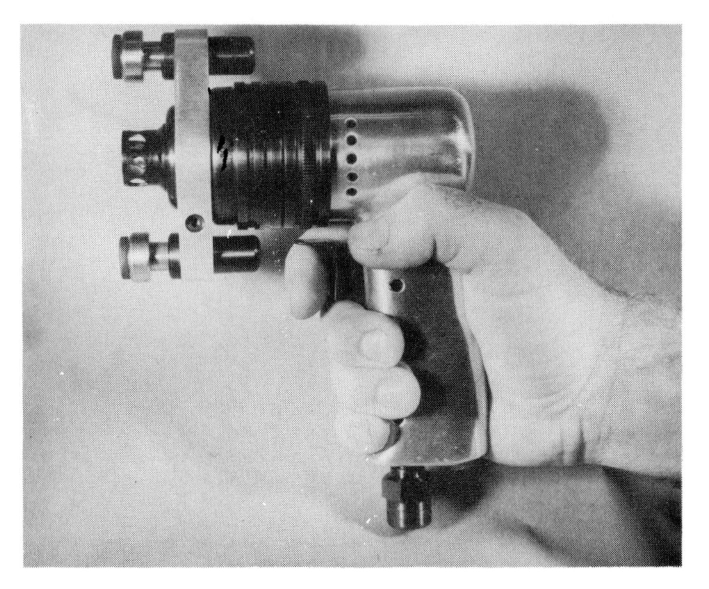

Fig. 10.—Rivet milling tool.

BLIND RIVETS

Every plane built has its own hard-to-get-at spots for the riveter. Some of these can be worked on from one side only, with no possible chance to use a bucking bar.

Blind riveting processes (driven from the head side only) have been developed to take care of those points inaccessible for rivet bucking which occur on all metal aircraft. Previous to the advent of the blind rivets, these points demanded an inferior conventional-type rivet, a bolt or screw, or in extreme cases, a design compromise to make the structure physically possible.

Explosive Rivet.—

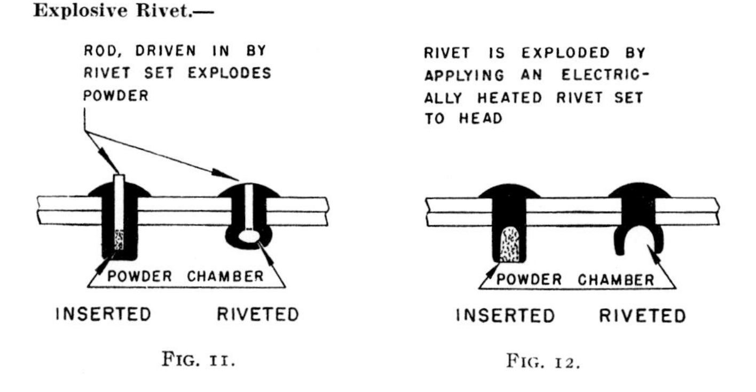

FIG. 11.

FIG. 12.

The first explosive-type rivet is drilled from the head side (Fig. 11), just deep enough to place the powder chamber past the grip. A small rod is then inserted to rest against the powder charge and protrude from the head slightly. After the rivet is inserted in the hole, a conventional-type rivet set is used to drive the rod into the chamber, exploding the powder and expanding the rivet.

The second explosive-type rivet is drilled from the shank end (Fig. 12) deep enough to place the powder chamber up to the grip. After the powder is injected, the hole is wadded with a plastic substance. To use, the rivet is placed in the hole drilled in the structure, then an electrically heated rivet set is applied to the head, exploding the powder charge and expanding the rivet.

Grip lengths should be checked very closely when using these rivets. Engineering approval usually is necessary.

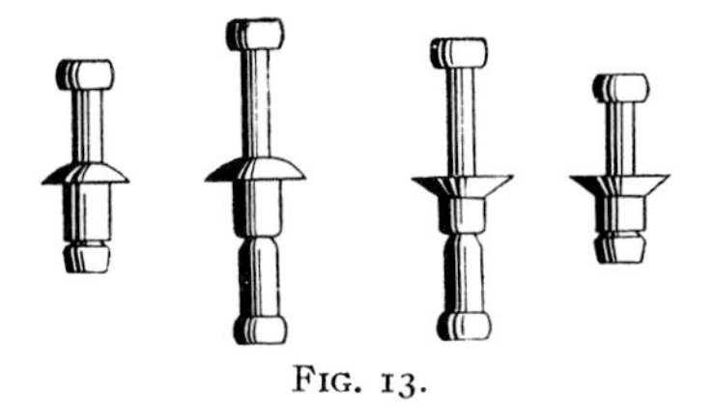

FIG. 13.

Cherry Blind Rivets are made of aluminum alloy and are exceedingly efficient. Their positive mechanical action leaves no doubt as to the formation of a satisfactory head on the blind side. They have high shear and fatigue values and do the job quickly.

Closing tanks, putting down floors, fastening inside skin on corrugated sections, applying closing plates to the underside of wings, and scores of other difficult or impossible jobs for conventional rivets are easy and fast with Cherry Blind Rivets.

There are four types of Cherry rivets (Fig. 13): from left to right, the hollow and self-plugging types with brazier head, the self-plugging and hollow types with countersunk heads. Each is available in four grip lengths and three diameters 1/8, 5/32 and 3/16 in.

The tools used to install mechanically expanded rivets depend upon the manufacturer of the rivet being installed. Each company has designed special tools which should always be used to ensure satisfactory results with its product. Hand tools as well as pneumatic tools are available.

After selection or determination of the rivet to be used in any installation, the proper size twist drill must be determined. (See Fig. 14)

Rivet size	Drill size
4 (1/8 in.)	#30
5 (5/32 in.)	#20
6 (6/32 in.)	#10
8 (1/4 in.)	F

Fig. 14

Finish drill sizes for common rivet shank diameters

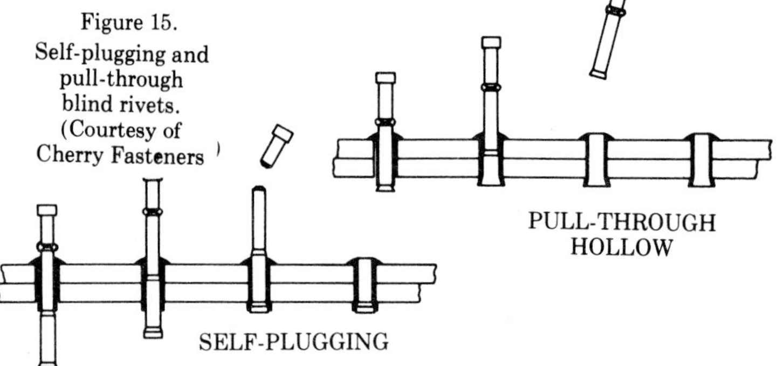

Figure 15. Self-plugging and pull-through blind rivets. (Courtesy of Cherry Fasteners)

Be very careful when drilling the material. Hold the drill at right angles to the work at all times to keep from drilling an elongated hole. The mechanically expanded rivet will not expand as much as a solid shank rivet. If the hole is too large or elongated, the shank will not properly fill the drilled hole. Common hand or pneumatic powered drills can be used to drill the holes. Some manufacturers recommend predrilling the holes; others do not.

Equipment used to pull the stem of the rivet, as previously stated, will depend upon the manufacturer of the rivet. Both manually operated and power-operated guns are manufactured for this purpose. Nomenclature for various tools and assemblies available depends upon the manufacturer. Application and use of the equipment is basically the same. Whether the equipment is called a hand tool, air tool, hand gun, or pneumatic gun (Fig. 16), all of these are used with but one goal, the proper installation of a rivet.

Pneumatic tools operate on the same air pressure as pneumatic riveting hammers, 90 to 100 p.s.i. Follow the operational procedures and adjustments recommended by the manufacturer.

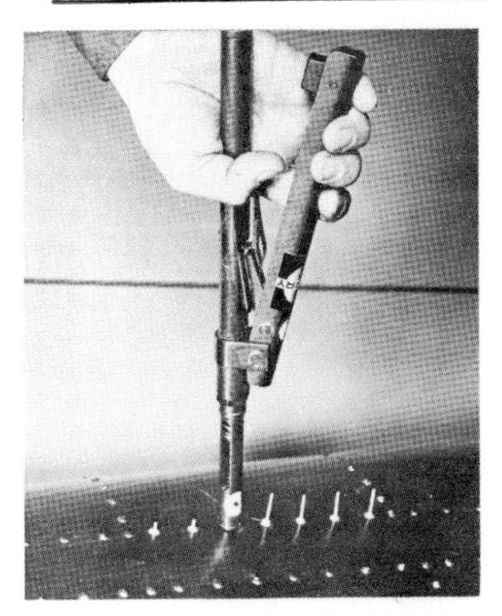

Figure 16. Tools used to install self-plugging (friction lock) rivets. *(Courtesy of Cherry Fasteners.)*

The choice of installation tools is influenced by several factors: The quality of rivets to be installed, the availability of an air supply, the accessibility of the work, and the size and type of rivet to be installed. In addition to a hand or power riveter, it is necessary to select the correct "pulling head" to complete the installation tool.

Selection of the proper pulling head is of primary importance since it compensates for the variables of head style and diameter. Since your selection will depend on the rivets to be installed, you should consult the applicable manufacturer's literature.

The inspection of installed mechanically expanded rivets is very limited. Often the only inspection that can be made is on the head of the rivet. It should fit tightly against the metal. The stem of the rivet should be trimmed flush with the head of the rivet whether it is a protruding head or a countersunk head.

HI-SHEAR RIVET

The Hi-shear rivet is made of heat treated steel and heat-treated aluminum alloys, and as its name implies, is designed for high shear strength. The rivet is not up-set in the usual manner as applied to rivets, but instead is held in place by an aluminum alloy ring which is swaged into a locking groove at the end of the rivet. Fig. 17

FIG. 17.

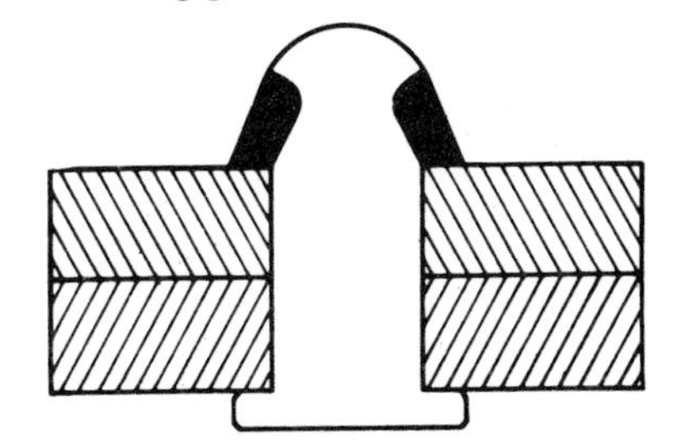

Fig. 18 shows progressive steps of locking the collar on the rivet. A special rivet set is used in the pneumatic rivet hammer to drive the collar while the rivet is backed up with a bucking bar. The riveting action first draws the collar tight against the material and as the driving action proceeds, swages the collar into the groove, and finally pinches off the excess of the collar. The amount of material sheared from the collar varies with the material thickness.

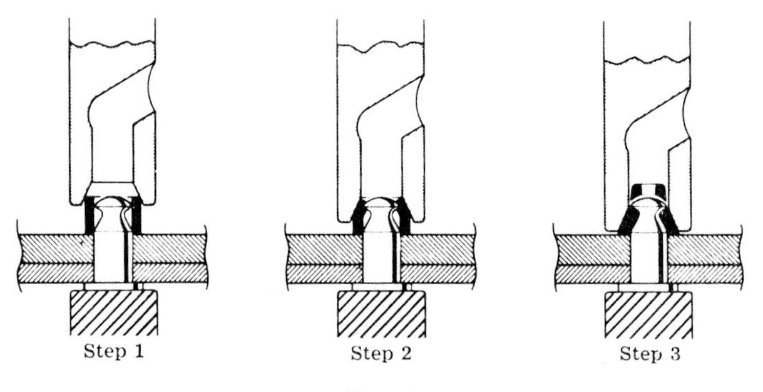

Fig. 18.

The rivets are made in increments of 16ths of an inch and to specify its length it is coded for the maximum material thickness (or grip) for which it is made. It is suitable for use on a material thickness one sixteenth less than this specified length.

The rivet is made with a brazier, flat-head, and countersunk head, and also in a close tolerance size.

The 100° countersunk and the flat-head steel rivet are coded NAS 177 and NAS 178 respectively. The locking collar of 2117 aluminum alloy is coded NAS 179. A dash number following the NAS number indicates the diameter in 32nds of an inch and a 2nd dash number indicates the maximum grip in 16ths of an inch. The other rivet head styles and materials mentioned above are coded by manufacturers' code.

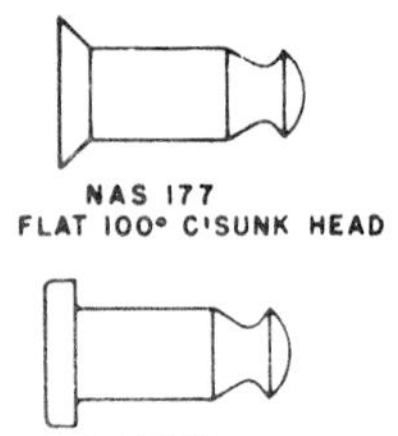

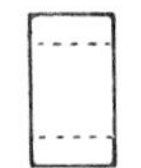

NAS 179 COLLAR

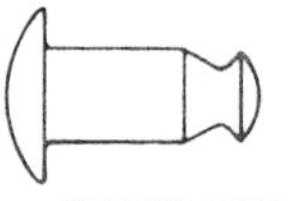

BRAZIER HEAD

Fig. 19.

CHAPTER II

BOLTS AND FASTENERS

BOLTING PRACTICE

Various types of fastening devices allow quick dismantling or replacement of aircraft parts that must be taken apart and put back together at frequent intervals. Bolts and screws are two types of fastening devices which give the required security of attachment and rigidity. Generally, bolts are used where great strength is required, and screws are used where strength is not the deciding factor.

The threaded end of a bolt usually has a nut screwed onto it to complete the assembly. The threaded end of a screw may fit into a female receptacle, or it may fit directly into the material being secured. A bolt has a fairly short threaded section and a comparatively long grip length or unthreaded portion, whereas a screw has a longer threaded section and may have no clearly defined grip length. A bolt assembly is generally tightened by turning the nut on the bolt; the head of the bolt may or may not be designed for turning. A screw is always tightened by turning its head.

Aircraft bolts are fabricated from cadmium- or zinc-plated steel, (usually nickel steel SAE2330), unplated corrosion-resistant steel, and anodized aluminum alloys. By far the most common is the cadmium plated steel bolt. Most bolts used in aircraft structures are either general-purpose, AN bolts, or NAS internal-wrenching or close-tolerance bolts, or MS bolts. In certain cases, aircraft manufacturers make bolts of different dimensions or greater strength than the standard types. Such bolts are made for a particular application, and it is of extreme importance to use like bolts in replacement. Special bolts are usually identified by the letter "S" or "spec" stamped on the head.

General-Purpose Bolts

The hex-head aircraft bolt (AN-3 through AN-20) is an all-purpose structural bolt used for general applications involving tension or shear loads where a light-drive fit is permissible. (.006-inch clearance for a ⅝-inch hole, and other sizes in proportion).

Alloy steel bolts smaller than No. 10-32 (3/16-inch diameter, AN3-) and aluminum alloy bolts smaller than ¼-inch diameter are not used in primary structures. Aluminum alloy bolts and nuts are not used where they will be repeatedly removed for purposes of maintenance and inspection.

The AN73-AN81 (MS20073-MS20074) drilled-head bolt is similar to the standard hex-bolt, but has a deeper head which is drilled to receive wire for safetying. The AN-3, AN-20 and the AN-73, AN-81 series bolts are interchangeable, for all practical purposes, from the standpoint of tension and shear strengths. See Chapter VIII Standard Parts.

Identification and Coding

Bolts are manufactured in many shapes and varieties. A clear-cut method of classification is difficult. Bolts can be identified by the shape of the head, method of securing, material used in fabrication, or the expected usage.

AN-type aircraft bolts can be identified by the code markings on the boltheads. The markings generally denote the bolt manufacturer, the material of which the bolt is made, and whether the bolt is a standard AN-type or a special-purpose bolt. **AN standard steel bolts are marked with either a raised dash or asterisk;** corrosion-resistant steel is indicated by a single raised dash; and AN aluminum alloy bolts are marked with two raised dashes. Additional information such as bolt diameter, bolt length, and grip length may be obtained from the bolt part number. See Chapter VIII, Standard Parts.

AIRCRAFT NUTS

Aircraft nuts are made in a variety of shapes and sizes. They are made of cadmium-plated carbon steel, stainless steel, or anodized 2024T aluminum alloy. No identifying marking or lettering appears on nuts. They can be identified only by the characteristic metallic luster or color of the aluminum, brass, or the insert when the nut is of the self-locking type. They can be further identified by their construction.

Aircraft nuts can be divided into two general groups: Non-self-locking and self-locking nuts. Non-self-locking nuts are those that must be safetied by external locking devices, such as cotter pins, safety wire, or locknuts. Self-locking nuts contain the locking feature as an integral part.

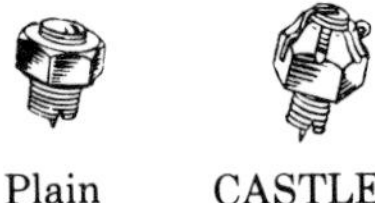

Plain AN 315 | CASTLE AN 310 | ELASTIC STOP NUT AN 365 MS 20365 | TINNERMAN | BOOTS SELF-LOCKING

FIG. 1.

Most of the familiar types of nuts including the plain nut, the castle nut, the castellated shear nut, the plain hex nut, the light hex nut, and the plain check nut are the non-self-locking type. (Fig. 1)

The castle nut, AN310, is used with drilled-shank AN hex head bolts, clevis bolts, eyebolts, drilled head bolts, or studs. It is fairly rugged and can withstand large tension loads. Slots (called castellations) in the nut are designed to accommodate a cotter pin or lock wire for safety. The AN310 castellated, cadmium plated steel nut is by far the most commonly used airframe nut. See Chapter VIII, Standard Parts.

The castellated shear nut, AN320, is designed for use with devices (such as drilled clevis bolts and threaded taper pins) which are normally subjected to shearing stress only. Like the castle nut, it is castellated for safetying. Note, however, that the nut is not as deep or as strong as the castle nut; also that the castellations are not as deep as those in the castle nut.

AIRCRAFT WASHERS

Aircraft washers used in airframe repair are either plain, lock, or special type washers.

Plain Washers

The plain washer, AN960 is used under hex nuts. It provides a smooth bearing surface and acts as a shim in obtaining correct grip length for a bolt and nut assembly. It is used to adjust the position of castellated nuts in respect to drilled cotter pin holes in bolts. Plain washers should be used under lockwashers to prevent damage to the surface material.

INSTALLATION OF NUTS AND BOLTS

Boltholes must be normal to the surface involved to provide full bearing surface for the bolthead and nut and must not be oversized or elongated. A bolt in such a hole will carry none of its shear load until parts have yielded or deformed enough to allow the bearing surface of the oversized hole to contact the bolt.

In cases of oversized or elongated holes in critical members, obtain advice from the aircraft or engine manufacturer before drilling or reaming the hole to take the next larger bolt. Usually, such factors as edge distance, clearance, or load factor must be considered. Oversized or elongated holes in noncritical members can usually be drilled or reamed to the next larger size.

Many boltholes, particularly those in primary connecting elements, have close tolerances. Generally, it is permissible to use the first lettered drill size larger than the normal bolt diameter, except where the AN hexagon bolts are used in lightdrive fit (reamed) applications and where NAS close-tolerance bolts or AN clevis bolts are used.

Light-drive fits for bolts (specified on the repair drawings as .0015-inch maximum clearance between bolt and hole) are required in places where bolts are used in repair, or where they are placed in the original structure.

The fit of holes and bolts is defined in terms of the friction between bolt and hole when sliding the bolt into place. A tight-drive fit, for example, is one in which a sharp blow of a 12- or 14-ounce hammer is required to move the bolt. A bolt that requires a hard blow and sounds tight is considered to fit too tightly. A light-drive fit is one in which a bolt will move when a hammer handle is held against its head and pressed by the weight of the body.

Examine the markings on the bolthead to determine that each bolt is of the correct material. It is of extreme importance to use like bolts in replacement. In every case, refer to the applicable Maintenance Instruction Manual and Illustrated Parts Breakdown.

Be sure that washers are used under both the heads of bolts and nuts unless their omission is specified. A washer guards against mechanical damage to the material being bolted and prevents corrosion of the structural members.

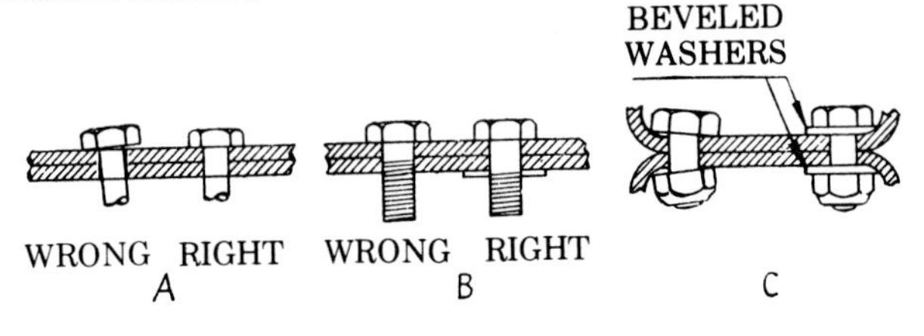

Fig. 2

Be certain that the bolt grip length is correct. Grip length is the length of the unthreaded portion of the bolt shank. See Figure 2. Generally speaking, the grip length should equal the thickness of the material being bolted together. However, bolts of slightly greater grip length may be used if washers are placed under the nut or the bolthead. In the case of plate nuts, add shims under the plate.

Self-locking nuts are not used with bolts that have a cotter pin hole in the threaded end. The sharp edge of the hole cuts the locking device and destroys its effectiveness.

A nut is not run to the bottom of the threads on the bolt (Fig. 3 B). A nut so installed cannot be pulled tight on the structure and probably will be twisted off while being tightened. A washer will keep the nut in the proper position on the bolt.

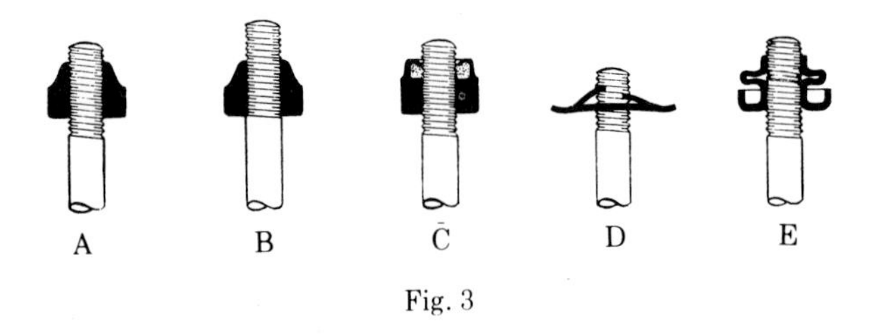

Fig. 3

In the case of self-locking "stop" nuts, if from one to three threads of the bolt extend through the nut, it is considered satisfactory.

Palnuts (AN356) should be tightened securely but not excessively. Finger-tight plus one to two turns is good practice, two turns being more generally used.

Torque Tables

The standard torque table should be used as a guide in tightening nuts, studs, bolts, and screws whenever specific torque values are not called out in maintenance procedures.

FINE THREAD SERIES		
BOLT SIZE	STANDARD TYPE NUTS (MS20365, AN310, AN315)	SHEAR TYPE NUTS (MS20364, AN320, AN316, AN23 THRU AN31)
10-32	20-25	12-15
1/4-28	50-70	30-40
5/16-24	100-140	60-85
3/8-24	160-190	95-110
7/16-20	450-500	270-300
1/2-20	480-690	290-410
9/16-18	800-1,000	480-600
5/8-18	1,100-1,300	660-780
3/4-16	2,300-2,500	1,300-1,500

COARSE THREAD SERIES		
BOLT SIZE	STANDARD TYPE NUTS (MS20365, AN310, AN315)	SHEAR TYPE NUTS (MS20364, AN320, AN316, AN23 THRU AN31)
8-32	12-15	7-9
10-24	20-25	12-15
1/4-20	40-50	25-30
5/16-18	80-90	48-55
3/8-16	160-185	95-110
7/16-14	235-255	140-155
1/2-13	400-480	240-290
9/16-12	500-700	300-420
5/8-11	700-900	420-540
3/4-10	1,150-1,600	700-950

Cotter Pin Hole Line-Up

When tightening castellated nuts on bolts, the cotter pin holes may not line up with the slots in the nuts for the range of recommended values. Except in cases of highly stressed engine parts, the nut may be over tightened to permit lining up the next slot with the cotter pin hole. The torque loads specified may be used for all unlubricated cadmium-plated steel nuts of the fine or coarse-thread series which have approximately equal number of threads and equal face bearing areas. These values do not apply where special torque requirements are specified in the maintenance manual.

If the head end, rather than the nut, must be turned in the tightening operation, maximum torque values may be increased by an amount equal to shank friction, provided the latter is first measured by a torque wrench.

Safetying of Nuts, Bolts and Screws

It is very important that all bolts or nuts, except the self-locking type, be safetied after installation. This prevents them from loosening in flight due to vibration.

Safety wiring is the most positive and satisfactory method of safetying capscrews, studs, nuts, boltheads, and turnbuckle barrels which cannot be safetied by any other practical means. It is a method of wiring together two or more units in such a manner that any tendency of one to loosen is counteracted by the tightening of the wire. See Fig. 4.

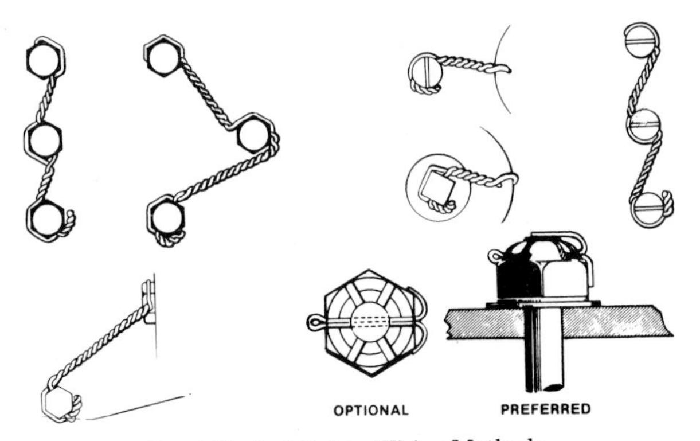

Fig. 4-Typical Safety Wiring Methods

COTTER PIN SAFETYING

Cotter pin installation is shown in Fig. 4. Castellated nuts are used with bolts that have been drilled for cotter pins. The cotter pin should fit neatly into the hole, with very little sideplay.

Machine screws, although used on the same general principle as bolts, require a different method of application. The threads usually run to the head and thus leave no grip for shear bearing. Machine screws, therefore, are used in tension with no concern for the threads extending into the hole.

The same rules hold for a screw as are illustrated in Fig. 1 for bolts.

A number of different types of heads are available on machine screws to satisfy the particular installation.

For any type of screw there is a correct screw driver. If the screw has a slotted head, the screw drioer should fit the slot snugly (Fig. 5A). The sides of the screw driver should, as nearly as possible, be parallel to the screw slot sides.

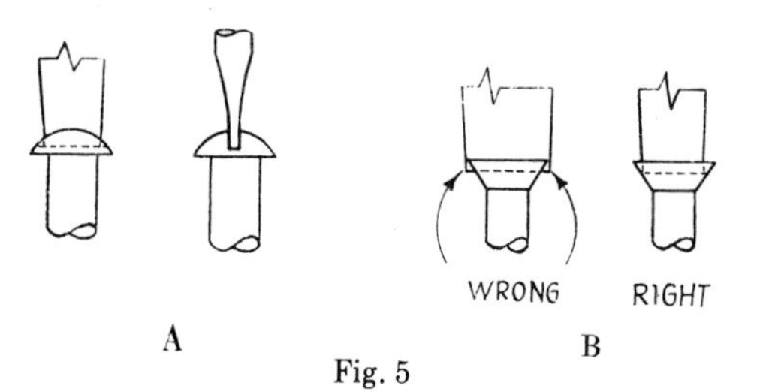

Fig. 5

Reed and Prince or Phillips heads require a special driver made for the particular screw. The drivers for the two are not interchangeable. (Fig. 6). The Phillips head has rounded shoulders in the recess while the Reed and Prince has sharp square shoulders. The use of the wrong screw driver on these screws may result in ruining the screw head. The use of power (electric and pneumatic) screw drivers has speeded up many installations, such as inspection doors and fillets, where the tool may be used in rapid succession on a row of screws.

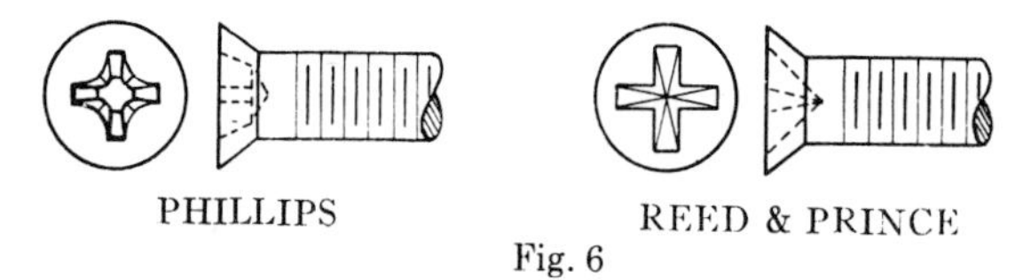

Fig. 6

Tapped Threads.—Instead of a nut, threads are often tapped into the bolted structure. In this case the bolts or screws are safetied with a wire through a hole drilled in the head (Fig 4). Whenever possible, the wire should be so strung that tension is held on the bolt or screw toward tightening it. Always keep in mind the fact that the wire should tend to tighten the screws.

Taper Pin Installation (using AN385 and AN386).—These pins, when properly installed, give the most serviceable splice method for torque

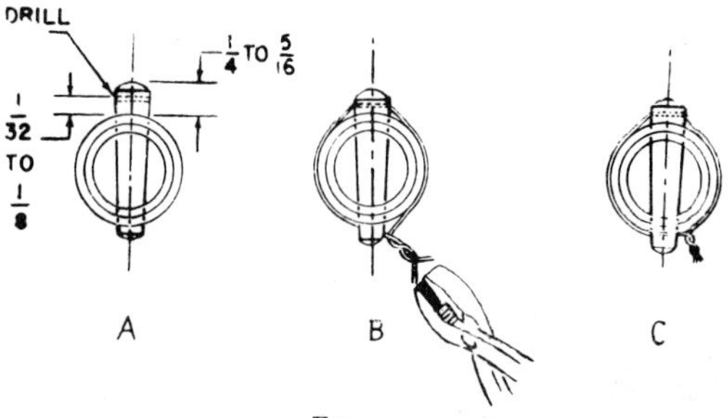

Fig. 7.

tube and rod. In the case of either pin, use the drill indicated on the blueprint or drawing; if the drill size is not indicated, use one from .003 to .005 smaller than the small end of the taper pin.

To install AN385, use a Morse taper reamer (¼ in. per foot) of the same designated size as the pin and ream a hole until the large end of the pin protrudes, as shown in Fig. 7A with the small end of the pin protruding slightly less. Drill No. 50 perpendicular to the axis of the torque rod, sizes 1 and larger; drill No. 60, sizes 4/0 to 1. Tap snugly in place, insert the safety wire, and pull the wire tight around the rod as it is twisted (Fig. 7C). Twist the wire only until it lies snugly against the tube or rod.

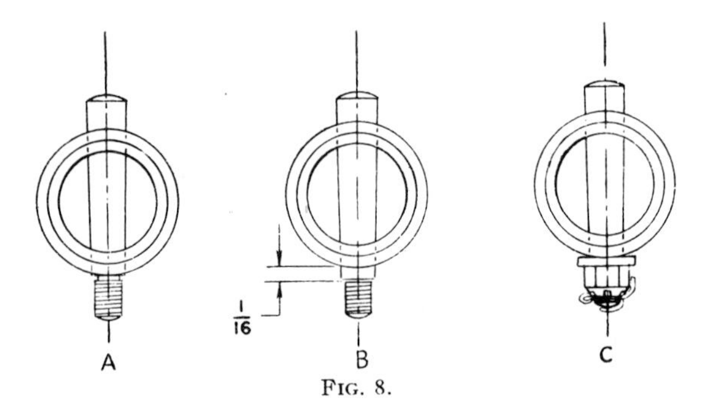

FIG. 8.

To install AN386 use a Brown and Sharpe taper reamer (½ in, per foot), same size as the indicated pin size, and ream a hole until the small end of the taper pin is flush to 1/16 in. over when tapped in place (Fig. 8A and B). Use AN975, same dash number as AN320 nut, with either AN320 nut or AN364 self-locking nut and pull the pin snugly into place. Cotter AN320 nut (Fig. 8C).

SPECIAL PURPOSE FASTENERS

The Elastic Stop Nut is a standard nut with the height increased to incorporate a fiber collar. This added height corresponds approximately to the thickness of a lock washer. The inside diameter of the collar is smaller than that of the bolt and is not threaded.

Before the bolt reaches the fiber washer, the Elastic Stop Nut acts in every way as a standard nut. The same play is present between the thread flanks as exists in standard nuts of comparable fit (Fig. 9).

Elastic Stop Nuts fit all standard bolts with fine or coarse thread from .06 of an inch in diameter for instruments to 6 in., or larger, for heavy machinery. The aircraft sizes are given in AN Standards AN 364, AN 365, AN 366, and AN 367.

When the bolt reaches the fiber collar, it tends to push the fiber up because the hole in the collar is smaller than the bolt and is not threaded (Fig. 10). This sets up a heavy downward pressure on the bolt, automatically throwing the load-carrying sides of the nut and bolt threads

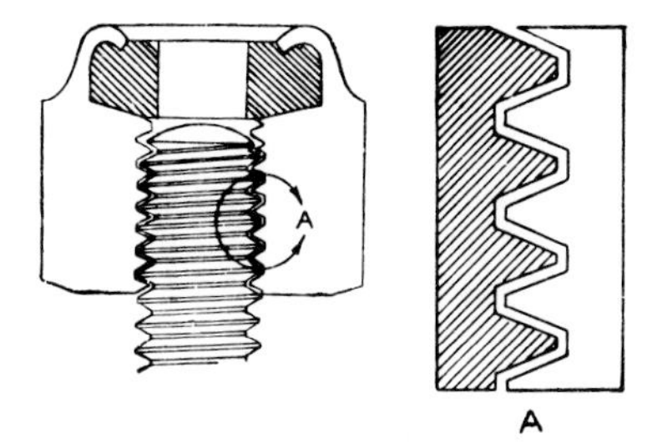

FIG. 9.

into positive contact. At once all play in the threads is eliminated and a friction is set up between every bolt and nut thread in contact.

After the thread pressure reaches a maximum, the bolt threads impress themselves into the fiber collar. They do not cut a thread.

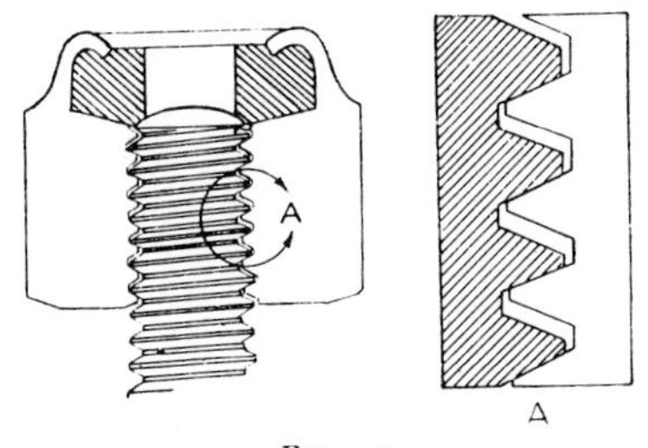

FIG. 10.

When the bolt has been forced all the way through the fiber collar, the downward pressure remains constant, eliminating the axial play and maintaining the required pressure on the load-carrying side of the bolt and nut threads (Fig. 11).

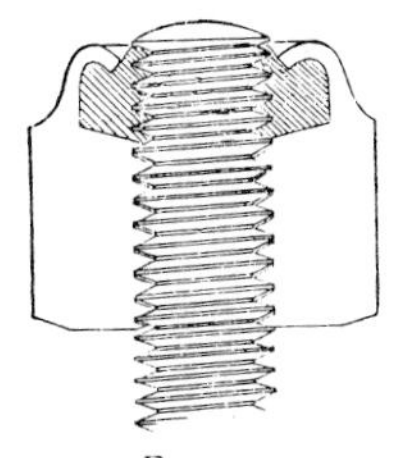

FIG. 11.

There is some frictional braking effect of the fiber on the bolt, but of greatest importance is the complete elimination of all axial play which is so detrimental to the conventional safetying methods such as cotter pins.

Standard practice in aircraft requires that at least one and not more than three threads extend through the nut and that no bolt with cotter-pin hole in shank be used with stop nuts: If more than three threads show, a washer should be used under the nut. A cotter-pin hole in the bolt will ruin the collar by cutting it away.

This unique self-locking nut has gained great popularity in the aircraft industry, there being as many as 30,000 used on one particular model of airplane. An Elastic Stop Nut may be reused many times without losing its effectiveness. The fiber collar seals the threads from outside weather conditions and thus prevents rust or corrosion from attacking the threads.

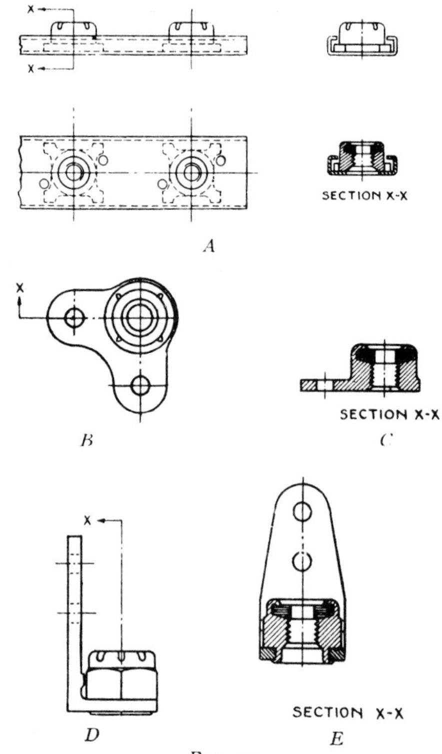

FIG. 12.

These nuts are made in a wide variety of shapes and from all materials applicable to the purpose or to match the screw or bolt. Odd shapes other than AN 364, AN 365, AN 366, and AN 376 are such as shown in Fig 12

The B. F. Goodrich Rivnut was originally designed for the attachment of deicers on the wing nose, where it is impossible to reach a nut on the inside. It is inserted and installed entirely from the outside. Because of its application, it has become quite popular as a means of attaching other parts and accessories where the situation prevents the use of conventional screws and bolts.

The Rivnut unit is composed of a Rivnut (Fig. 13) and a screw (Fig. 14), either attachment or plug type. Rivnuts are manufactured in three types (Fig. 13): *A*, flathead; *B*, countersunk 115 deg; and *C*, countersunk 100 deg. There are two head styles in the 100-deg Rivnut; one with head thickness of .063 in. (same thickness as 115 deg), and one of .048 in. for machined countersink in thinner material.

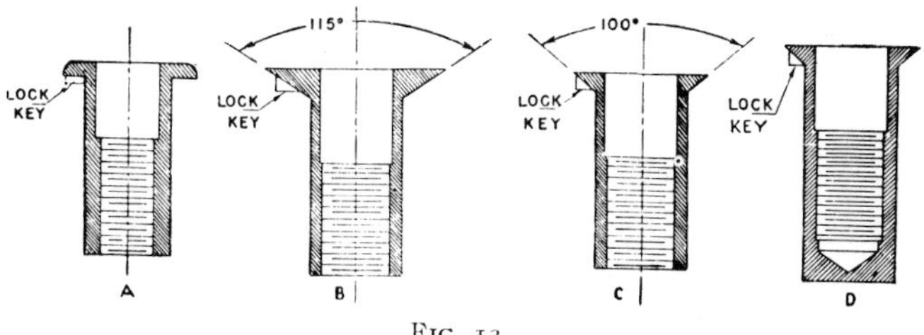

FIG. 13.

All these types are also manufactured in a closed shank type (Fig. 13*D*). The closed type is indicated in code number by B and is intended for use in a sealed compartment.

Rivnuts may be obtained with or without a key. Use of the key is indicated by a K in code number. K precedes B when the two letters are both present in the code.

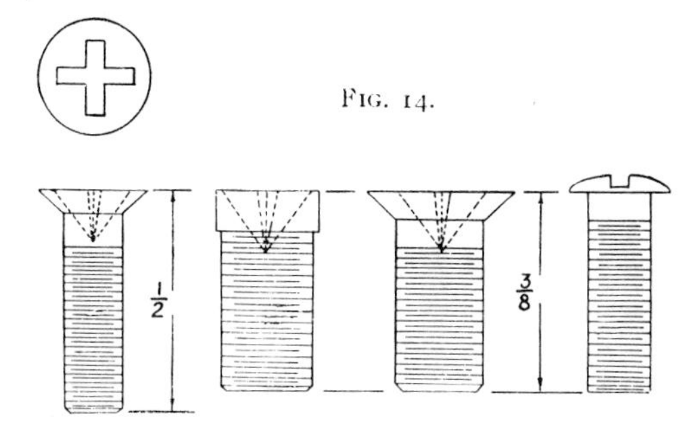

FIG. 14.

Each type is manufactured in three sizes: 6–32, 8–32, and 10–32, having outside diameters of 3/16, 7/32, and 1/4 in., respectively.

The part number consists of the screw size, as noted above (6, 8, and 10), and the maximum grip length by the following table. Maximum grip in thousandths of an inch is indicated by figures at the right of the type number. Minimum grip is maximum grip of next smaller size. Minimum grip of shortest size is head thickness for countersunk types and .010 for flathead type. The length of the grip is also indicated by marks on the head, as illustrated above the following table, the shortest having no mark.

Type and head thickness	(no mark)	(1 mark)	(2 marks)	(3 marks)	(4 marks)	(5 marks)
Flathead (.032)	.045	.075	.100	.120	.140	100
100° countersunk (.048)	.091	.121	.146	.166	.186	.206
100° countersunk (.063)	.106	.136	.161	.181	.201	.221
115° countersunk	.107	.137	.162	.182	.202	.222

Examples of part number:

6KB45 Rivnut is flathead, size 6-32, closed end, with key and .045 maximum grip, and .010 in. minimum grip.

8B106 Rivnut is 100-deg countersunk head (.063 thick), size 8-32, closed end, with .106 maximum grip and .063 in. minimum grip.

10-107 Rivnut is 115-deg countersunk head (.063 thick), size 10-32, open end, with .107 maximum grip and .063 in. minimum grip.

As noted above, two types of screws are used with Rivnuts (Fig. 14): plug screws 3/8 in. long, made of 1015 steel, cadmium-plated, or 2024 aluminim alloy anodized, come in headless, countersunk, or brazier head with Reed & Prince type, or Phillips type, recess. Attachment screws are ½ in. long countersunk head, with Reed & Prince type, or Phillips type, recess, made of S.A.E. steel heat-treat hardened and cadmium-plated.

To install the Rivnut: drill a hole of proper size, cut a key recess with B.F. Goodrich key-seat cutter (C-3376), if key type is used, and insert

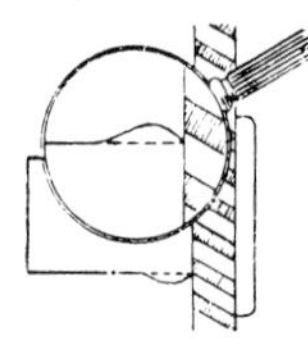

INSUFFICIENT UPSET

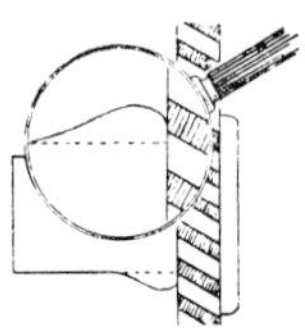

CORRECT UPSET

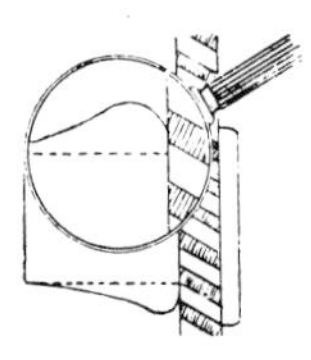

EXCESSIVE UPSET

Fig. 15.

the Rivnut with any of the following tools: Chicago pneumatic CP301R Rivnut driver, or Goodrich hand heading tools. Pull the Rivnut down to give a good uniform upset, as shown in Fig. 15, and remove the tool by unscrewing from the Rivnut.

Tinnerman Speed Nuts for aircraft apply the spring tension locking principle (in many different shapes and sizes for various attachments) in aircraft manufacture. This principle has been successfully used in other industries such as the automotive, refrigerator, stove, and radio, for some time.

Note (Fig. 16 *A*) the starting position, how the arched prongs fit into the threads, while the main base of the Speed Nut is well arched. As the bolt is turned and tightened, the main arch of the Speed Nut is brought down and the prongs are forced deeper into the roots of the threads to the double-locked position (Fig. 16 *B*). This gives an arched spring lock and an inward thread lock at the same time.

Speed Nuts are normally approved for use on non-structural attachments.

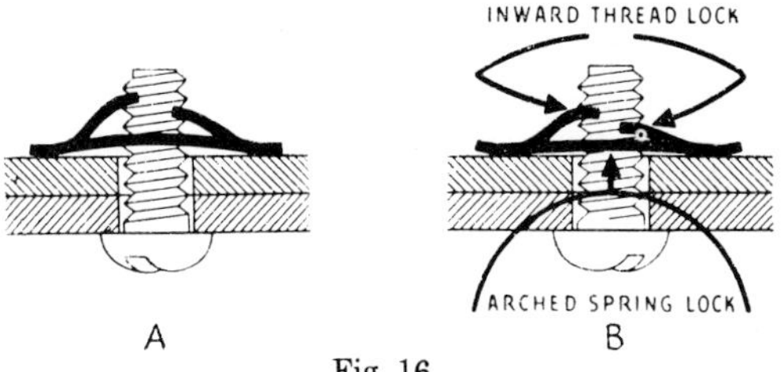

Fig. 16

Standard Speed Nuts are made of special aircraft spring steel and are also available in stainless steel for spot welding in screw-receiving position. They are now being furnished with a protective finish of Parkerizing, followed by zinc chromate primer. Where Speed Nuts might function as an electrical connector, they are furnished cadmium-plated.

Standard Speed Nuts fit AN515 screws in sizes 4-40, 6-32, 8-32, 10-24, and ¼-20. AN530 sheet-metal screws in sizes 4Z, 6Z, 8Z, 10Z, and 14Z.

Speed Nuts cost and weigh considerably less and their free-running fit makes them extremely fast to apply. Moreover, being almost flat, they do not require so long a screw as other nuts, so space is conserved and weight is reduced still more through the use of shorter and lighter screws.

Fig. 17.

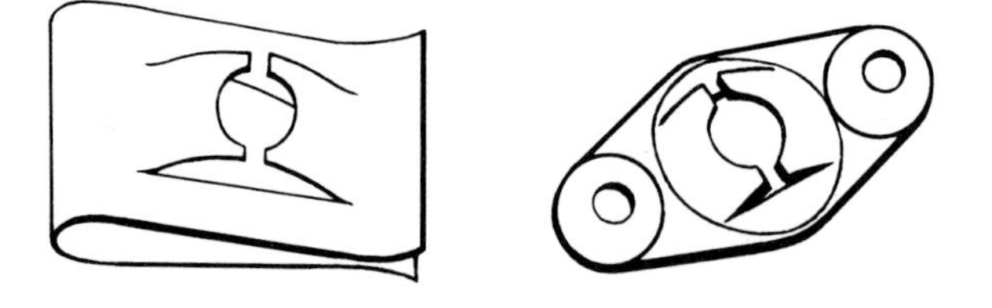

Fig. 18.

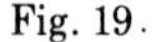

Fig. 19.

Twin-type Speed Nuts (Fig. 17) are ideal for attaching split fair-lead guide blocks or other attachments where fasteners are grouped in pairs. The use of a wrench to hold the nut is unnecessary.

U-type Speed Nuts (Fig. 18) are slipped over the edges of panels or sheets and hold themselves in screw-receiving position. They eliminate the welding or riveting of anchor-type lock nuts in location.

Double-lug anchor-type Speed Nuts (Fig. 19) are available for regular or flush blind mounting, for attachments having plain screw and rivet holes, or for dimpled holes in sheets to accommodate 100-deg countersunk head screws and rivets; also for spot welding.

Dill Lok-Skru Fastener.—This part is widely useful for blind attachments of all kinds. Since it is installed by one man from one side, it answers many difficult construction problems. The Lock-Skru (Fig. 20) is composed of three parts: the barrel A, the head B, and the attachment screw C.

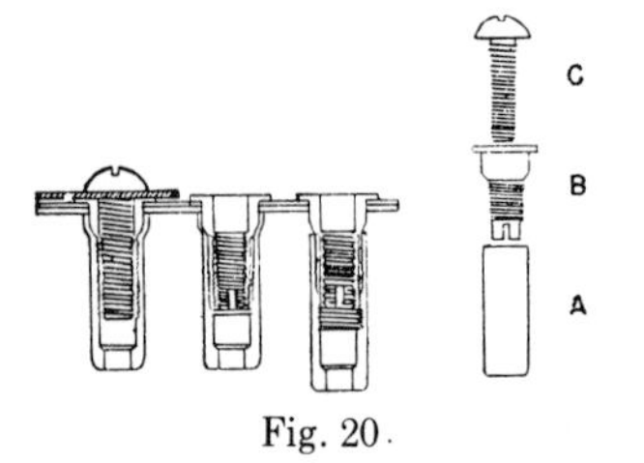

Fig. 20.

The following installation directions are for the hand tool, although speedier production is possible with the power tool. These steps are standard practice with most aircraft manufacturers (Fig. 21).

1. The tool is inserted in the Lok-Skru so that the blade extends through the barrel slot and the driver is firmly set in the head slot. The Lok-Skru is then inserted in the drilled hole.

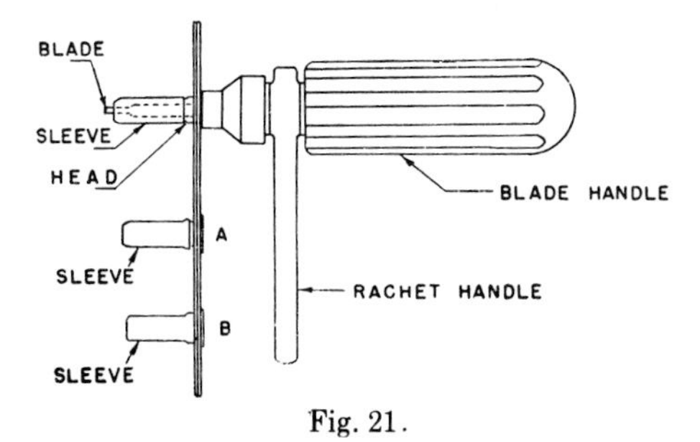

Fig. 21.

2. With the ratchet handle held stationary, the barrel blade handle is turned to the left until the sleeve has come up firmly against the sheet on the other side. The tool is pressed firmly in the Lok-Skru to hold the tool blade and the driver in the slots.

3. The barrel handle should not be turned after the Lok-Skru barrel has been drawn against the sheet as in Fig. 21*B*. Final tightening should be made by taking a quarter turn or less on the ratchet handle, thus drawing the head into the sheet. The blade handle should be held stationary while turning the ratchet handle.

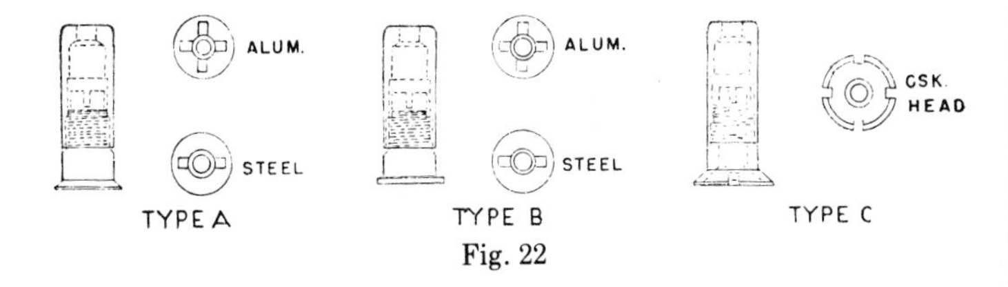

Fig. 22

4. After a tightness test of installation with a small (8 in.) screw driver, filed round on the end, the Lok-Skru is ready for attachments.

Lok-Skrus are made in five different types and from either steel or aluminum alloy. Types *A*, *B*, and *C* are shown in Fig. 22, types *D* and *E* in Fig. 23.

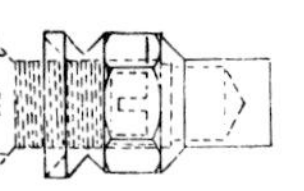

TYPE D

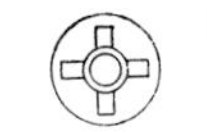

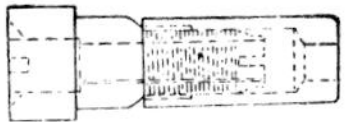

TYPE E

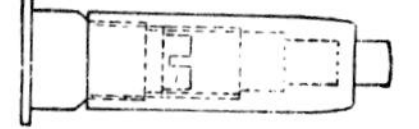

CLOSED END BARRELS

Fig. 23.

TYPES, DRILL SIZES, AND MATERIALS OF LOK-SKRUS

Type	Hole size	Head metal	Screw size	Type	Hole size	Head metal	Screw size
A	.297	Steel	10–32	*C*	.297	Steel	10–32
A	.297	Aluminum	8–32	*C*	.265	Steel	8–32
A	.297	Aluminum	10–24	*C*	.265	Aluminum	6–32
A	.265	Steel	8–32	*C*	.234	Steel	6–32
A	.265	Aluminum	6–32				
A	.234	Steel	6–32	*D*	.250	Steel	8–32
				D	.281	Steel	10–32
B	.297	Steel	10–32	*D*	.250	Aluminum	6–32
B	.297	Aluminum	8–32	*D*	.281	Aluminum	8–32
B	.297	Aluminum	10–24	*D*	.250	Chamet Bronze	8–32
B	.265	Steel	8–32				
B	.265	Aluminum	6–32				
B	.234	Steel	6–32	*E*	.297	Aluminum	10–32
				E	.297	Aluminum	8–32
				E	.265	Aluminum	6–32

LOK-RIVETS
(Similar to Lok-Skrus but not tapped)

Type	Hole size	Metal head	Type	Hole size	Metal head
A	.265	Steel	*B*	.265	Steel
A	.265	Aluminum	*B*	.265	Aluminum
A	.234	Steel	*B*	.234	Steel
A	.234	Aluminum	*B*	.234	Aluminum
A	.187	Steel	*B*	.187	Steel
A	.187	Aluminum	*B*	.187	Aluminum

The type D hexagonal Lok-Skru is made especially for fuel tanks and other tight compartments (Fig. 23).

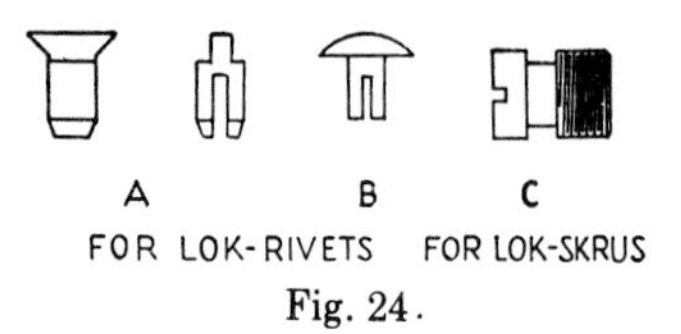

Fig. 24.

Figure 24 shows plug screws and plugs for both Lok-Skrus and Lok-Rivets.

Dzus Fasteners

The Dzus turnlock fastener consists of a stud, grommet, and receptacle. Figure 25 illustrates an installed Dzus fastener and the various parts.

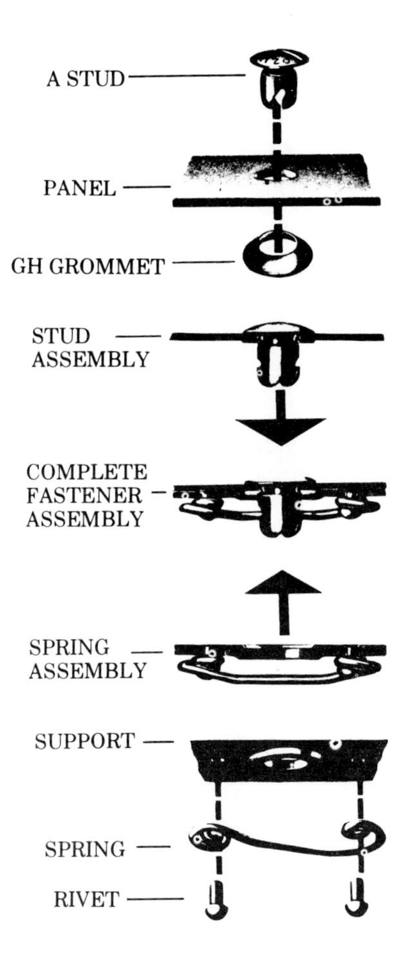

Figure 25
The Dzus fastener.
(Dzus Fastener Co. Inc. illustration)

The grommet is made of aluminum or aluminum alloy material. It acts as a holding device for the stud. Grommets can be fabricated from 1100 aluminum tubing, if none are available from normal sources.

The spring is made of steel, cadmium plated to prevent corrosion. The spring supplies the force that locks or secures the stud in place when two assemblies are joined.

The studs are fabricated from steel and are cadmium plated. They are available in three head styles; wing, flush, and oval.

A quarter of a turn of the stud (clockwise) locks the fastener. The fastener may be unlocked only by turning the stud counter-clockwise. A Dzus key or a specially ground screwdriver locks or unlocks the fastener.

Special installation tools and instructions are available from the manufacturers.

CHAPTER III

TOOLS AND THEIR PROPER USE

DRILLS AND DRILLING

Drilling.—The drill is a tool for originating and enlarging holes in metal or other substances. The twist drill, usually employed for general drilling purposes, has the following parts (Fig. 1):

1. *Point.*—The cone-shaped surface on the cutting end.
2. *Shank.*—The part of the drill that fits into the drill-press spindle or drill chuck.
3. *Body.*—The part from the point to the shank.
4. *Dead Center.*—The extreme end of the drill point.
5. *Heel.*—That portion of the drill point back of the tips or cutting edges

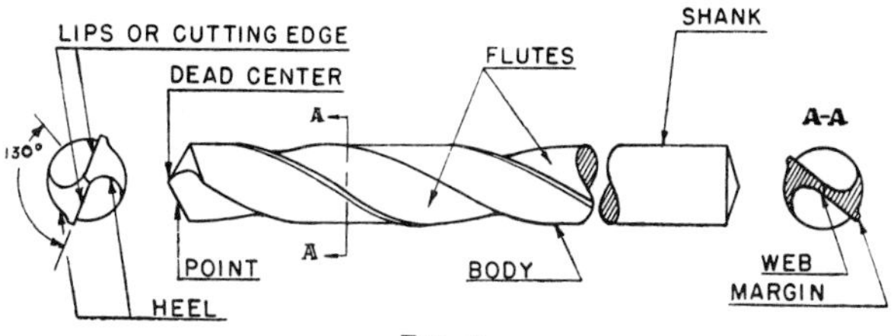

Fig. 1.

6. *Flutes.*—Two spiral grooves cut in the body of the drill to permit free-cutting lips and the removal of the chips from the hole.
7. *Lips.*—The cutting edges of the drill point.
8. *Tang.*—The small end of a tapered shank drill. (Note the *margin* that gives body clearance.)
9. *Web.*—The supporting section of the body of the drill lying between the flutes.

Drill sizes are designated by letters (from A to Z), by fractions giving the diameter, and by numbers giving wire drill size. See the table (page 143) for a list of drill sizes and their decimal equivalents.

Drill Grinding.—It is judged that fully 95 per cent of the difficulties encountered in drilling are caused by faulty grinding. A few pointers on the process follow:

1. The lengths of the cutting edges should be equal, otherwise the drill will cut an oversize hole and is liable to break.

2. The angle between the cutting edges should be approximately 118 or 59 deg from the drill axis. A smaller (to 90 deg) included angle (Fig. 3*A*) should be used for soft materials such as aluminum, Bakelite, lead, or wood and a larger (to 150 deg) included angle (Fig. 3*B*) for hard materials such as steels—stainless, manganese, or heat-treated. For brass, the lips of a blunt or wide-angle point are ground straight with the drill axis as shown in (Fig. 3*C*). This is to prevent "hogging in" as the standard drill point will do.

3. The correct lip clearance should be as illustrated in order that the heel will not drag and thus prevent the cutting edge from feeding properly. (Fig. 2*B*).

4. The dead center of the web should meet the cutting edges correctly as shown in Fig. 1.

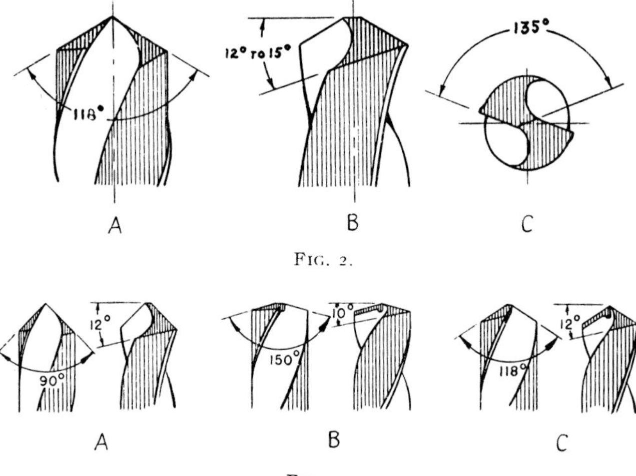

FIG. 2.

FIG. 3.

Drilling.—The handling of a drill motor, though simple, should be carefully done as illustrated in Fig 4*A*

The drill motor is held securely by both hands. The thumb and fingers of one hand are extended to contact the work in order to steady the drill in starting as well as act as a safety stop to prevent the drill from damaging anything back of the skin.

Figure 4*B* shows what can happen if control of the drill is not kept at all times. The stringer is certain to be damaged and may have to be replaced. It is well to investigate, before drilling, where the drill point is and what it is apt to hit. When in doubt, stop and look.

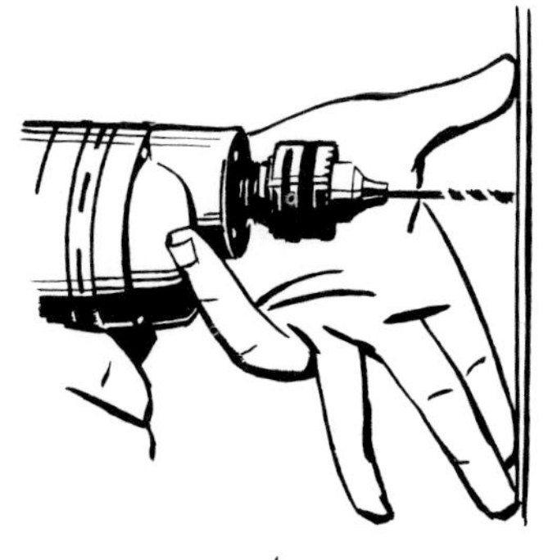

A

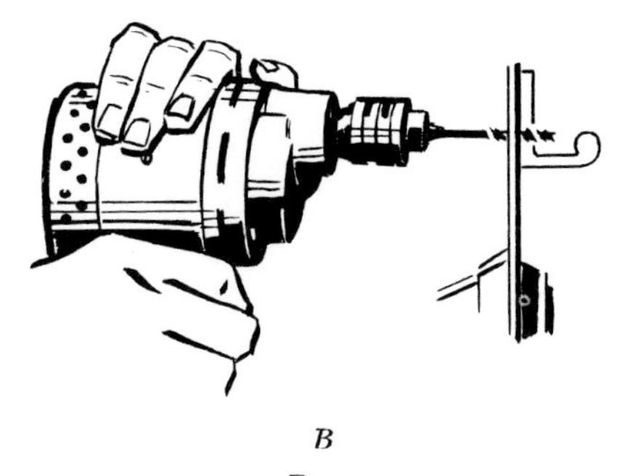

B

FIG. 4.

In addition to the conventional electric drill motor, the pneumatic drill motor or "air drill" is becoming extremely popular, because the operator can vary the speed from high to slow to suit the particular situation.

When drilling above the eye level goggles are used to protect the eyes against chips.

The workman who knows his drills, sharpens them correctly, and uses them carefully with the correct feed, speed, and lubricant, finds drilling a pleasant task.

Drill speeds are, of course, dependent upon many factors, but particularly upon the material being drilled. The following information is considered good practice and is for reference in case of doubt. The speed and feed combination is very important and, although the suggestions listed below are a helpful guide, the good mechanic watches his drill point and chips to judge a job. If the outer corners of the point wear away rapidly or burn, too much speed is indicated; if the drill chips along the cutting edge, too much feed is indicated. In the case of a power feed, .007 to .015 in per revolution (the smaller the drill, the smaller the feed) is good; in the case of a hand feed or a hand-held drill motor, sufficient pressure should be exerted to keep a steady uniform chip coming. To figure drill speeds by the following tables, check the recommended cutting speed for the given material in the Cutting Speeds table, then check this speed along the top of the Drill Speeds table against the drill size in the left-hand column.

CUTTING SPEEDS*

Material	Feet per minute	Material	Feet per minute
Aluminum	150–300	Tool steel	60–70
Brass	70–100	Stainless steel	30–40
Cast iron	70–100		

* These speeds for high-speed drills should be cut in half for carbon drills.

DRILL SPEEDS

Drill size, in.	Feet per minute				Drill size, in.	Feet per minute			
	30	70	100	150		30	70	100	150
	Revolutions per minute					Revolutions per minute			
1/16	1,833	4,278	6,111		7/16	262	611	873	1,310
1/8	917	2,139	3,056	4,584	1/2	229	535	764	1,146
3/16	611	1,426	2,037	3,056	5/8	183	428	611	917
1/4	458	1,070	1,528	2,292	3/4	153	357	509	764
5/16	367	856	1,222	1,833	7/8	131	306	437	655
3/8	306	713	1,019	1,528	1	115	267	382	573

Lubricant.—Drilling efficiency is greatly increased by the use of the proper drill lubricant, which serves as a coolant as well. The following table is suggested as good practice.

LUBRICANT

Metal	Drilling	Reaming	Tapping threading
Steels:			
Machinery	1	2	1 or 2
Carbon tool	1 or 2	2	2 or 4
High speed	1 or 2	2	2 or 4
Forgings	1 or 2	2	2 or 4
Stainless	2	2	2
Monel	2	2	2
Iron:			
Malleable	1	1	1
Cast	6	6	6
Aluminum	2, 3, or 6	2, 3, or 6	1 or 3
Brass	2 or 3	1	1 or 2
Copper	1	1	1 or 2
Bronze	1 or 6	1 or 6	1 or 2

DRILL LUBRICANTS

1. Soluble oil.
2. Lard oil.
3. Kerosene.
4. Sulphur-base oil.
5. Turpentine.
6. Dry (no lubricant recommended).

NOTE: The Cleveland Twist Drill Company publishes the most authoritative information on drilling in its "Handbook for Drillers."

Countersinking and counterboring, or spot-facing, are closely related to drilling operations in that each is applied to a drilled hole.

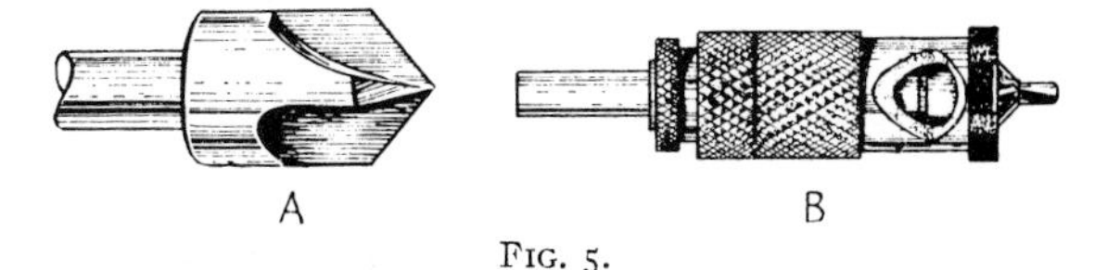

FIG. 5.

A rivet or screw hole is countersunk to allow the rivet- or screwhead to lie flush with the surface. The countersink (Fig. 5) may be operated by a hand drill, by a portable electric drill motor, or by a drill press.

When the conventional type of countersink (Fig. 5*A*) is used, it is necessary to try each hole with a rivet or screw as a precaution against getting it sunk too deep. When it is desirable to use a power tool, the stop countersink (Fig. 5*B*) is best because the depth of countersink can be controlled. In either case, the countersink shaft should be held perpendicularly to the surface of the work (Fig. 6*C*). Otherwise an unsatisfactory off-center hole will result and leave the edge of the hole exposed on one side and the screw edge rising on the other, as in Fig. 6*B*. In Fig. 6*A* the hole and the screwhead are concentric.

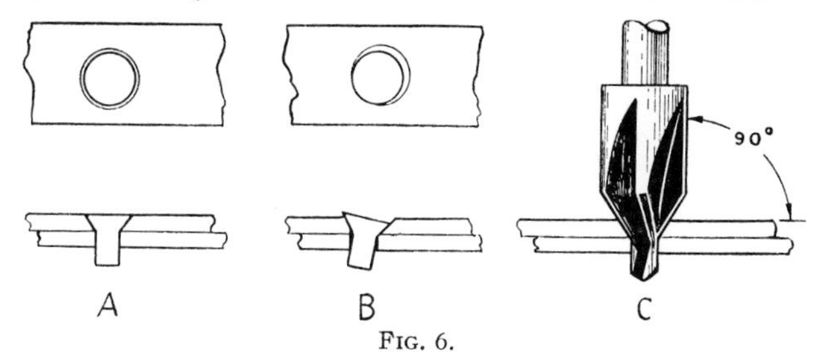

FIG. 6.

A counterboring tool is like a bottoming drill with a pilot and is used to sink a hole concentric with another to allow the bolt or screw-head to be sunk below the surface (Fig. 7). Since the action of this tool simulates that of a milling machine cutter, extreme care should be taken to have a snugly fitting pilot and a steadily held power supply.

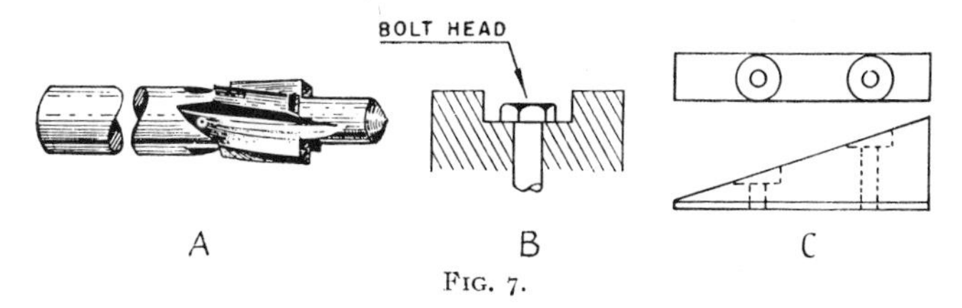

FIG. 7.

The spot-facer is a counterboring tool generally used in small drill presses and hand drill motors. Its application in aircraft structure is principally to provide a surface for a bolt or screwhead when the hole is necessarily drilled off perpendicular to the natural surface of the casting, forging, or extrusion (Fig. 7C).

FILES AND FILING

Files are classified in three ways: by name, according to shape or style; by type, according to the type cut; and by grade, according to the grade of the cut.

The names and shapes of a few of the most popular files are:

Hand—taper width, parallel thickness.

Mill—taper width, parallel thickness.

Pillar—taper thickness, parallel width.

Warding—much taper width, parallel thickness.

Square, round and *three square*—taper.

Half round—taper.

Knife—taper.

Vixen—parallel edges and sides.

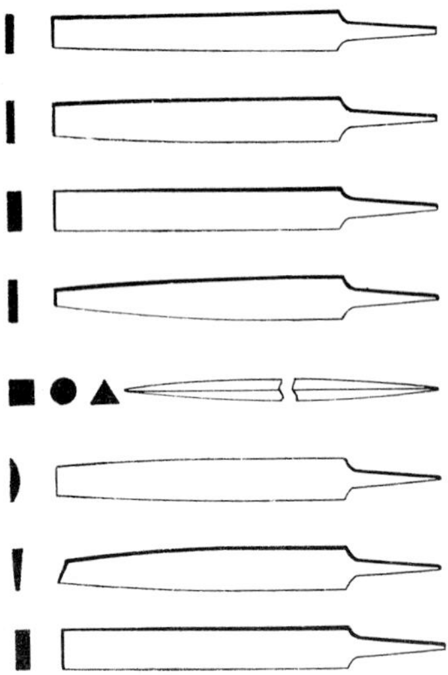

Fig. 8.

American Swiss and Vixen (Fig. 9 *C* and *D*) have special cuts and are intended for special work. The Swiss type is very fine double cut at 45-deg angle, making it very fast cutting. The Vixen, as illustrated, uses a single, knifelike, curved scraper cut which cleans well as it cuts. Swiss files also come in many special shapes for die sinking, etc.

The types of general-purpose files are single cut, with a single row of parallel teeth running across the face, and double cut with two rows of parallel teeth crossing at an angle (Fig. 9 *A* and *B*).

The principal grades are as follows:

No. 000 cut.............	Coarse cut	No. 2 and 3 cuts...	Smooth cut
No. 00 cut..............	Bastard cut	No. 4 and 5 cuts...	Dead smooth cut
No. 0 and 1 cuts........	Second cut		

Swiss files use only the numerical expression of the cut. Vixen files use only three grades: regular, fine, and smooth. The length of a file does not include the tang. Some files, such as the hand and the pillar, have one or two safe (smooth) edges.

The important points in filing are as follows: use a straight, forward motion with steady firm pressure; prevent the file from "rocking" over

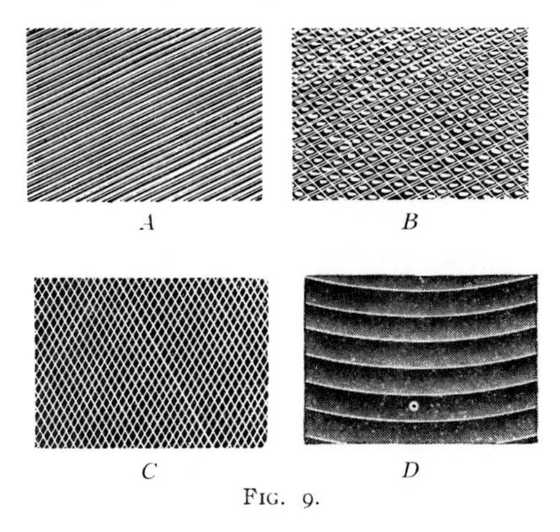

Fig. 9.

the work; avoid dragging the file back across the work, as this will soon knock the edge off the file; keep the file clean by tapping it against the bench or by brushing it with a file card or wire brush.

MEASURING TOOLS

Rules are graduated measuring instruments usually made of metal or wood. Flexible steel rules are the most commonly used in aircraft work. Figure 10 shows the common 6-in. flexible steel rule graduated in 10ths and 100ths on one side and 32nds and 64ths on the opposite.

Scales, although similar to rules in appearance, are graduated to indicate proportional rather than actual measurements, as, for example, in laying out work one quarter size. Rules and scales have straight edges and may be used as a straightedge although it is considered better practice to use a straightedge designed for that purpose if one is available.

Fig. 10.—(*Courtesy of Lufkin Rule Co.*)

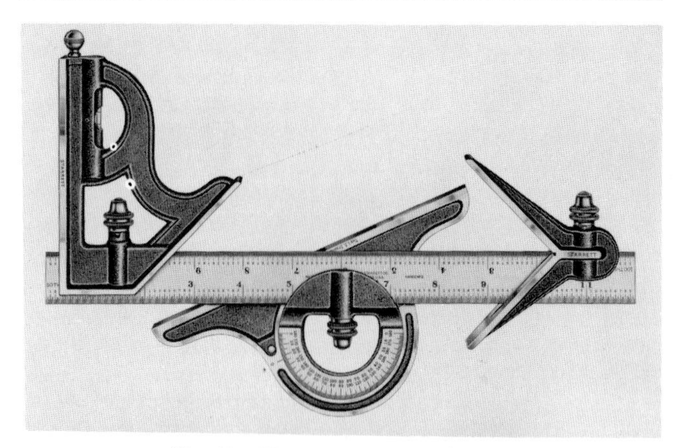

Fig. 11—*(Courtesy of L.S. Starrell Co.)*

The combination square, so called because it is a combination of a rule and various angle heads, is probably the most popular all-round measuring tool. (Fig. 11).

Calipers, the most popular precision measuring tools, are used for comparing or measuring distances. The types shown in Fig. 12 are outside calipers and inside calipers and are used for comparing or transferring measurements.

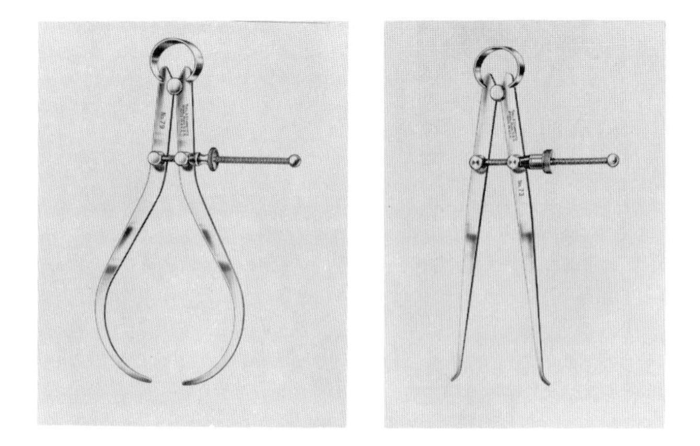

Fig. 12—*(Courtesy of L.S. Starrell Co.)*

Micrometer calipers are calipers graduated for precision measuring down to .001 or .0001 in. There are three types of micrometers: inside, outside, and depth. A thread micrometer is a special type of outside instrument for measuring the pitch diameter of threads.

In general aircraft work, the 1-in. outside micrometer is the most common and is therefore used as the example here. It is composed of the major parts as shown in Fig. 13. It actually records the endwise travel of a screw during a whole turn or any part of a turn. The screw of a micrometer has a pitch of 40 threads to the inch; in other words, if the screw is turned 40 times, it will move the spindle exactly 1 in. either toward or away from the anvil. Therefore, a single turn of the screw moves the spindle 1/40 in. or 25 thousandths of an inch.

The graduated inch on the barrel is divided into 40 parts, each representing one turn of the thimble, or 25 thousandths of an inch. Each fourth division is numbered 1 through 10, representing 100 thousandths, 200 thousandths, etc. (Fig. 13).

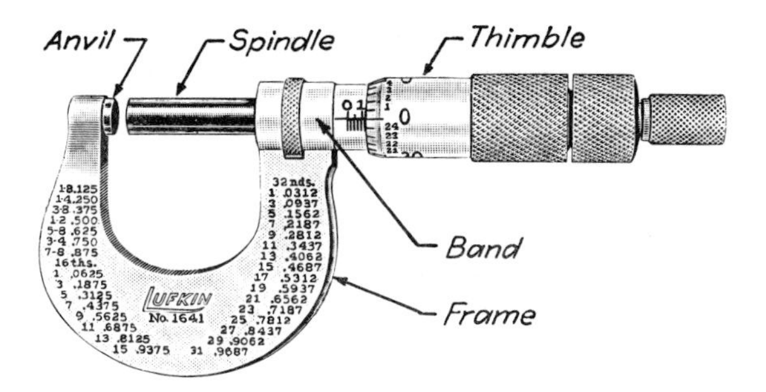

FIG. 13.—(*Courtesy of Lufkin Rule Co.*)

Now if the spindle moves 25 thousandths for every revolution of the thimble, then 1/25 of a revolution must move the spindle one thousandth of an inch. Note then, the numbers 0 to 25 running around the thimble (Fig. 13), each representing one thousandth of the inch. Read the one aligning with the horizontal line on the barrel.

The micrometer reading consists of the total of the 25 thousandths lines visible plus the reading on the thimble. Thus the micrometer in Fig. 13 reads 125 thousandths (5 lines times 25 thousandths)

Figure 14 illustrates three examples of readings. First read the micrometer illustration yourself, then check with the reading following the figure. Figure 14A reads .304. Figure 14B reads .226. Figure 14C reads .224. Note the handy decimal equivalents on the frame of the Lufkin micrometer in Fig. 13.

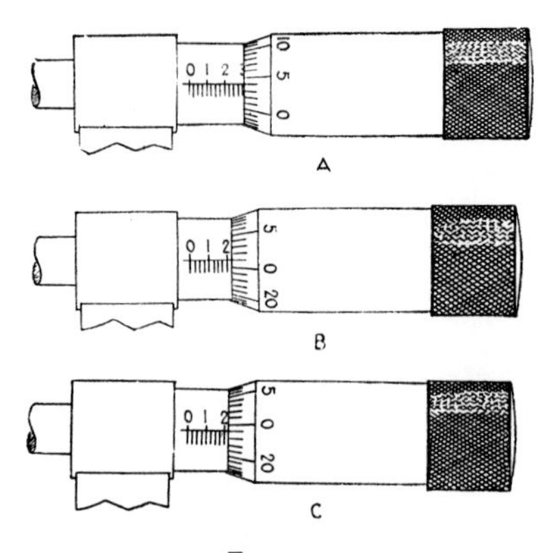

FIG. 14.

TAPS AND DIES

Taps are tools used for cutting inside or female threads in holes in metal, fiber, or other material. The two standard types are standard hand taps and machine screw taps.

Standard hand taps are made for cutting threads from ¼ in. in diameter and larger, while machine screw tap sizes are from 0 to 10. Both are designated by the screw size and threads.

Taps are in three forms: taper tap for starting; plug tap for use when the bottom of the hole is closed; bottoming tap for cutting threads clear to the bottom of the hole. Fig. 15.

It is very important to start the tap straight and keep it so throughout the work, because taps, especially small ones, are easily broken if strained. Use a lubricant, page 46. Start the tap with steady even pressure. As soon as the tap is started, it will feed itself though. Turn the tap a half turn clockwise, then back up a quarter turn to allow it to clean.

Dies are used for cutting outside or male threads. Start the large side of the die on the work and press down firmly. As soon as the die starts, it will feed itself. Use the same motion as for taps, one half turn forward and reverse one quarter turn. Use the proper lubricant for the material.

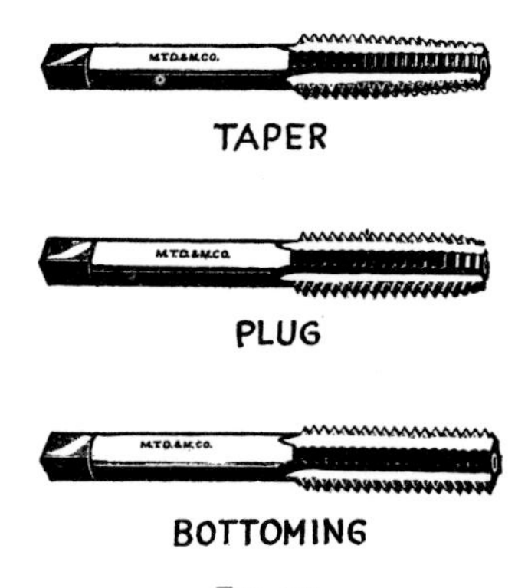

FIG. 15.

REAMERS

Reamers are tools for enlarging drilled holes to an exact diameter while finishing them round, straight, and smooth. The two types in general use are straight reamers and taper reamers.

There are three types of straight reamers: the solid reamer, one solid piece of steel; the expansion reamer, slotted with a tapered screw to expand or contract it a few thousandths of an inch; the adjustable reamer with blade separate from the body, which is tapered with lock nuts to move the blades to larger or smaller size. The solid reamer, being ground to size and therefore more accurate, is used mostly in production; the other two are used chiefly for service and repair.

Taper reamers, as the name implies, are used to ream holes for taper pins or bolts. They are of standard taper either ¼ or ½ in. per foot. Be sure that the reamer is started straight with the hole and that the hole is the right size. Turn the reamer clockwise (never counter-clockwise, even when removing) with just enough pressure to keep it feeding. A reamer should be kept absolutely clean to do good work. Never allow it to strike other tools or lie against them. Always keep the blades covered with a thin film of oil to prevent rust.

CHAPTER IV

ASSEMBLY AND INSTALLATION METHODS

AIRCRAFT PLUMBING

Aircraft plumbing is, as the name implies, the system of tubing used for fuel, oil, coolant, hydraulic, instrument, and vent lines and carries the "life blood" of the aircraft. These lines are coded with colored bands to facilitate tracing and to avoid confusion on assembly or installation.

Seamless tubing is used exclusively for aircraft plumbing and all lines, excepting vent lines, are given a pressure test to eliminate any chance of leakage.

Materials used in the tubing for aircraft plumbing are copper, aluminum alloy, stainless steel (18-8), and Inconel. Tubing sizes are designated by the outside diameter.

Fittings are made of aluminum alloy, brass and steel. The illustration shows two types of fittings joined to the tube end. The nut and sleeve are placed on the tube prior to the flaring of the tube. With the tube

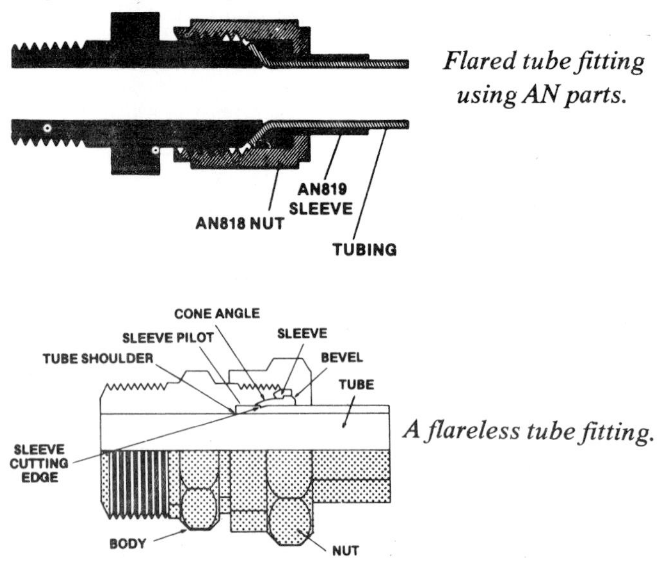

Flared tube fitting using AN parts.

A flareless tube fitting.

Fig. 1—Plumbing connectors

then flared, tightening the nut clamps the tube flare between the sleeve and the fitting.

The two fittings shown in Fig. 1 are similar but are different in design.

The first (an AN fitting) is made specifically to AN sizes and specifications. The complete group of AN fittings are identified individually by AN numbers and are illustrated in the "Standard Parts" section—AN774 through AN928.

The flaring tool used for aircraft tubing has male and female dies ground to produce a flare of 35 degrees to 37 degrees. Under no circumstances is it permissible to use an automotive type flare tool which produces a flare of 45 degrees.

The MS (Military Standard) flareless-tube fittings are finding wide application in aircraft plumbing systems. Using this type fitting eliminates all tube flaring, yet provides a safe, strong, dependable tube connection. The fitting consists of three parts: a body, a sleeve, and a nut. The body has a counterbored shoulder, against which the end of the tube rests. (See Figure 1) The angle of the counterbore causes the cutting edge of the sleeve to cut into the outside of the tube when the two are joined.

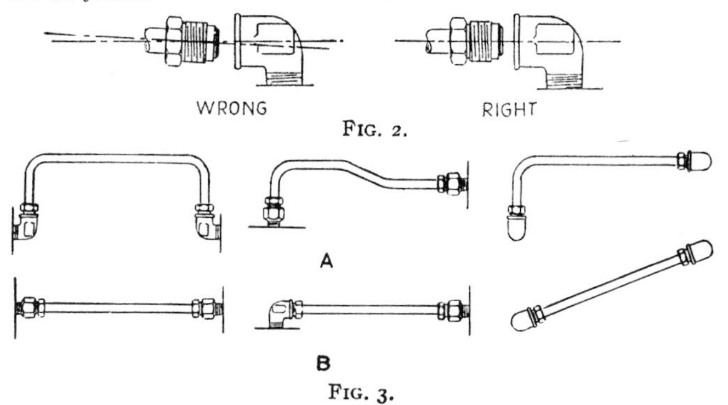

FIG. 2.

FIG. 3.

Installation—When tubing assemblies are being installed, the fittings should be properly aligned. With the nuts slipped back, the flares should sit easily in place and require no springing of the tubes to make them seat. The central axis of the tube and fitting should align properly to avoid cross threading (Fig. 2). Lines should be laid out with offsets as shown in Fig. 3A, avoiding point-to-point lines as shown in Fig. 3B.

Some compound, such as Parker Threadlube, is used on all threads **except on oxygen plumbing.** *Never use grease on oxygen lines. For oxygen fittings with straight threads, use Aquadag and for oxygen fittings with pipe threads, use litharge and glycerin.* Threadlube is used sparingly, as the surplus entering the lines will cause trouble. Nuts are run down by hand, then tightened with smooth jaw wrenches, one on the nut and one on the fitting, according to the accompanying torque table.

TORQUE VALUES FOR TIGHTENING FLARED TUBE FITTINGS

SIZE	-4	-6	-8	-10	-12	-16	-24
TUBE O.D.	1/4	3/8	1/2	5/8	3/4	1	1 1/2
Torque inch-lbs Alum Alloy Tubing, Fitting or Nut	40-65	75-125	150-250	200-350	300-500	500-700	600-900
Steel Tubing, Fitting or Nut	135-150	270-300	450-500	650-700	900-1000	1200-1400	1500-1800
Hose End Fittings, MS28740 End Fittings	100-250	200-480	500-850	700-1150			

Flexible hose is used in aircraft plumbing to connect moving parts with stationary parts in locations subject to vibration or where a great amount of flexibility is needed. It can also serve as a connector in metal tubing systems.

Synthetic materials most commonly used in the manufacture of flexible hose are: Buna-N, Neoprene, Butyl and Teflon (trademark of DuPont Corp.). **Buna-N** is a synthetic rubber compound which has excellent resistance to petroleum products. Do not use for phosphate ester base hydraulic fluid (Skyrol®). **Neoprene** is a synthetic rubber compound which has an acetylene base. Its resistance to petroleum products is not as good as Buna-N but has better abrasive resistance. Do not use for phosphate ester base hydraulic fluid (Skydrol®). **Butyl** is a synthetic rubber compound made from petroleum raw materials. It is an excellent material to use with phosphate ester based hydraulic fluid (Skydrol®). Do not use with petroleum products. **Teflon** is the DuPont trade name for tetrafluoroethylene resin. It has a broad operating temperature range (-65 degrees F. to +450 degrees F.). It is compatible with nearly every substance or agent used.

Flexible rubber hose consists of a seamless synthetic rubber inner tube covered with layers of cotton braid and wire braid, and an outer layer of rubber-impregnated cotton braid. This type of hose is suitable for use in fuel, oil, coolant, and hydraulic systems. The types of hose are normally classified by the amount of pressure they are designed to withstand under normal operating conditions.

1. Low pressure, any pressure below 250 p.s.i. Fabric braid reinforcement.
2. Medium pressure, pressures up to 3,000 p.s.i.
   One wire braid reinforcement.
   Smaller sizes carry pressure up to 3,000 p.s.i.
   Larger sizes carry pressure up to 1,500 p.s.i.

3. High pressure (all sizes up to 3,000 p.s.i. operating pressures).

Identification markings consisting of lines, letters, and numbers are printed on the hose. These code markings show such information as hose size, manufacturer, date of manufacture, and pressure and temperature limits. Code markings assist in replacing a hose with one of the same specification or a recommended substitute. Hose suitable for use with phosphate ester base hydraulic fluid will be marked "Skydrol® use." In some instances several types of hose may be suitable for the same use. Therefore, in order to make the correct hose selection, always refer to the maintenance or parts manual for the particular airplane.

The size of flexible hose is determined by its inside diameter. Sizes are in one-sixteenth-inch increments and are identical to corresponding sizes of rigid tubing, with which it can be used.

Support clamps are used to secure the various lines to the airframe or powerplant assemblies. The rubber-cushioned clamp is used to secure lines subject to vibration; the cushioning prevents chafing of the tubing. The plain clamp is used to secure lines in areas not subject to vibration.

A Teflon-cushioned clamp is used in areas where the deteriorating effect of Skydrol® 500, hydraulic fluid (MIL-0-5606), or fuel is expected. However, because it is less resilient, it does not provide as good a vibration-damping effect as other cushion materials. Use bonded clamps to secure metal hydraulic, fuel, and oil lines in place. Unbonded clamps should be used only for securing wiring.

AIRCRAFT ELECTRICAL SYSTEMS

The satisfactory performance of any modern aircraft depends to a very great degree on the continuing reliability of electrical systems and subsystems. Improperly or carelessly installed wiring or improperly or carelessly maintained wiring can be a source of both immediate and potential danger. The continued proper-performance of electrical systems depends on the knowledge and techniques of the mechanic who installs, inspects, and maintains the electrical system wires and cables.

Procedures and practices outlined in this section are general recommendations and are not intended to replace the manufacturer's instructions and approved practices.

The wires and cables are assembled in groups wherever possible, in order that they can be run through a protective metal conduit.

The multiple-contact electrical connector (AN3100, 3102, 3106, and 3108) has made the assembly and service of airplanes, in production units, much easier. These connectors are popularly referred to as "Cannon plugs" after their manufacturer, the Cannon Electric Development Company.

Electrical connectors are available with almost any number of contacts, from 1 to 100 as required.

Some wire manufacturers color-code the wires for size.

Wire lengths for electrical assemblies are predetermined, so that the electricians are able to cut them and make up the assembly at the bench without fitting the wires in the airplane.

The wires are next fitted with terminals or connectors on one end and laid out on templates or pin boards in order that the terminal on the other end will be attached in exactly the right position. The board or template duplicates the length over which the wires will be strung in the ship. The loose ends are held in one convenient position for attaching the fittings.

After the wires are complete with terminals, they are grouped together and laced into a cable. This lacing consists of knotting one end of small stout cord around one end of the group of wires, then with a series of locking double half hitches spaced every 1½ in., the wires are snugly grouped. This cable or group of wires is strung easily in a conduit and is less likely to be damaged than if it were left loose until installed.

Conduit is either solid or flexible. The solid conduit is made up much the same as plumbing except that electrical conduit fittings are used with thin-walled tubing. Flexible conduit is a woven-wire tube equipped with attaching fittings. It is popularly referred to as "Breeze" after the manufacturer, The Breeze Company.

More recently the transparent plastic Surco tubing has been equipped with attaching fittings and used in more protected places as conduit. This tubing is also extensively used over a laced cable which is to be strung through a previously installed conduit. These installations are made by tying a string to one end of the assembly, blowing the other end through the installed conduit, then pulling the assembly through. While the string is being blown through, the other end of the tube should be kept away from the faces and eyes of other workers.

Copper Wire Terminals Copper wires are terminated with solderless, preinsulated straight copper terminal lugs. The insulation is part of the terminal lug and extends beyond its barrel so that it will cover a portion of the wire insulation, making the use of an insulation sleeve unnecessary (figure 4).

In addition, preinsulated terminal lugs contain an insulation grip (a metal reinforcing sleeve) beneath the insulation for extra gripping strength on the wire insulation. Preinsulated terminals accommodate more than one size of wire; the insulation is usually color-coded to iden-

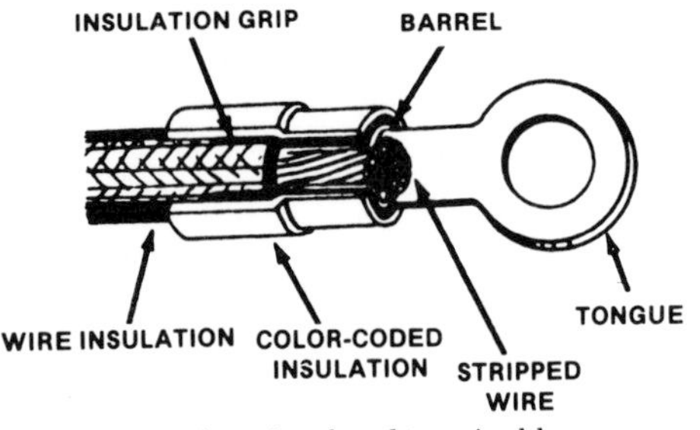

Figure 4. A preinsulated terminal lug.

tify the wire sizes that can be terminated with each of the terminal lug sizes.

Hand, portable power, and stationary power tools are available for crimping terminal lugs. These tools crimp the barrel of the terminal lug to the conductor and simultaneously crimp the insulation grip to the wire insulation.

Hand, crimping tools all have a self-locking ratchet that prevents opening the tool until the crimp is complete. Some hand crimping tools are equipped with a nest of various size inserts to fit different size terminal lugs. Others are used on one terminal lug size only. All types of hand crimping tools are checked by gages for proper adjustment of crimping jaws. Figure 5 shows a terminal lug inserted into a hand tool.

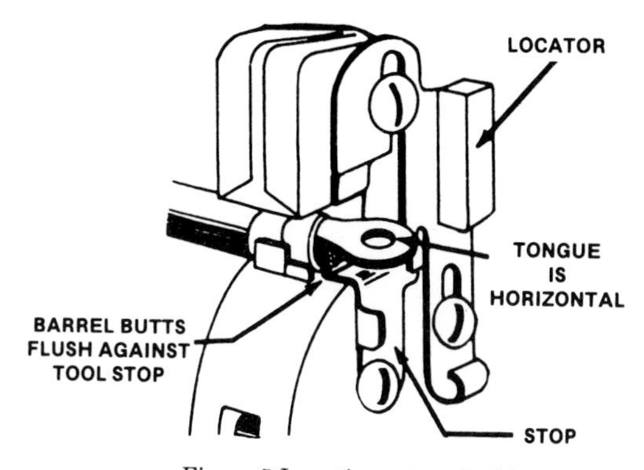

Figure 5 Inserting a terminal lug into a hand crimping tool.

Some types of uninsulated terminal lugs are insulated after assembly to a wire by means of pieces of transparent flexible tubing called "sleeves." The sleeve provides electrical and mechanical protection at the connection. When the size of the sleeving used is such that it will fit tightly over the terminal lug, the sleeving need not be tied; otherwise, it should be tied with lacing cord as illustrated in figure 6.

Aluminum Wire Terminals

The use of aluminum wire in aircraft systems is increasing because of its weight advantage over copper. However, bending aluminum will cause "work hardening" of the metal, making it brittle. This results in failure or breakage of strands much sooner than in a similar case with copper wire. Aluminum also forms a high-resistant oxide film immediately upon exposure to air. To compensate for these disadvantages, it is important to use the most reliable installation procedures.

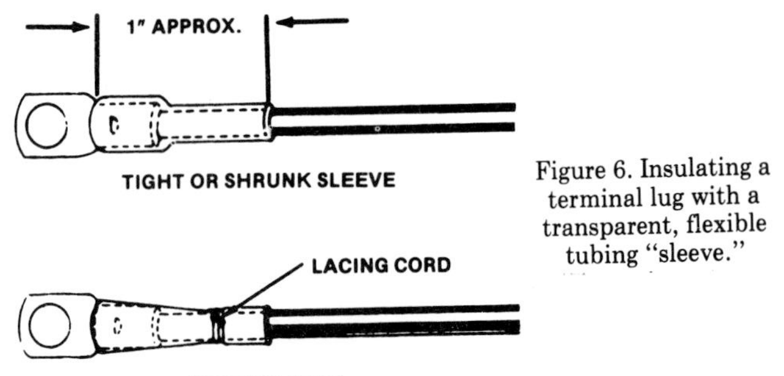

Figure 6. Insulating a terminal lug with a transparent, flexible tubing "sleeve."

Only aluminum terminal lugs are used to terminate aluminum wires. They are generally available in three types: (1) Straight, (2) right-angle, and (3) flat. All aluminum terminals incorporate an inspection hole (figure 7) which permits checking the depth of wire insertion. The barrel of aluminum terminal lugs is filled with a petrolatum-zinc dust compound. This compound removes the oxide film from the aluminum by a grinding process during the crimping operation. The compound will also

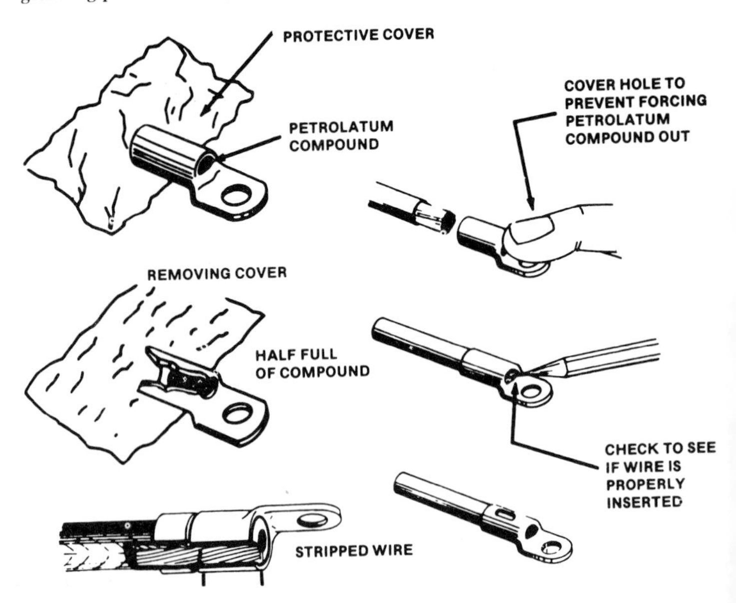

Figure 7. Inserting aluminum wire into aluminum terminal lugs.

minimize later oxidation of the completed connection by excluding moisture and air. The compound is retained inside the terminal lug barrel by a plastic or foil seal at the end of the barrel.

AIRPLANE CONTROL SYSTEMS

Airplane control mechanisms transmit motion from the pilot's controls in the cockpit to the proper control surface or accessory, using torque tubes, push-pull tubes, and cable-pulley arrangements.

Torque tubes, as the name implies, transmit the load by a turning motion with push-pull tubes or cable-operated bell cranks on either end. Torque tubes are mounted on roller or ball bearings. When torque tubes are being installed, the mounting brackets should be

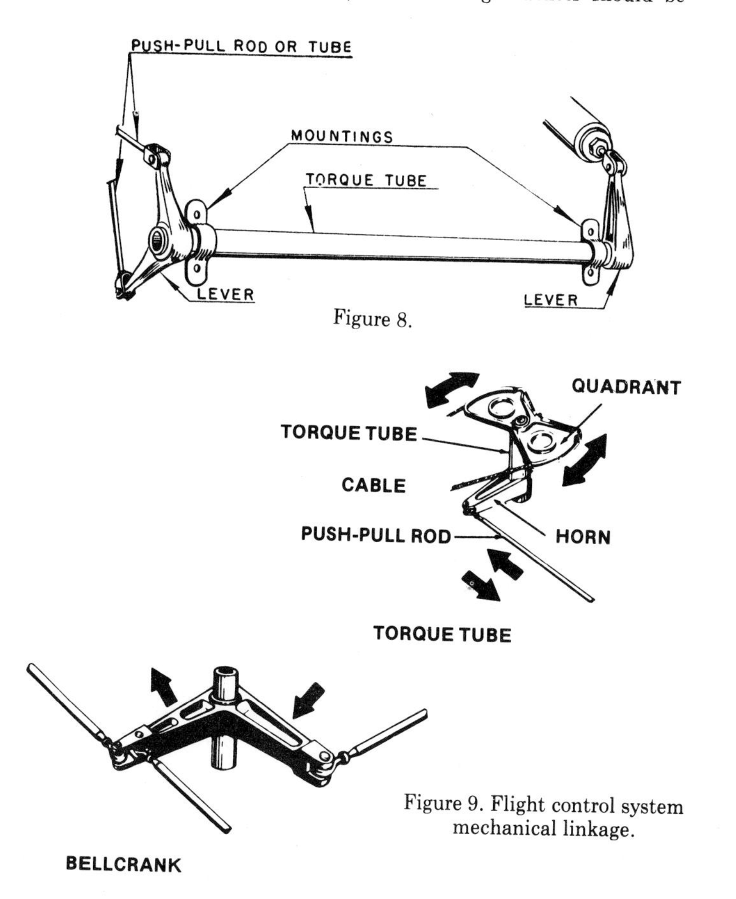

Figure 8.

Figure 9. Flight control system mechanical linkage.

properly aligned so that they will not bind the bearings when bolts are tightened. Levers or bell cranks are riveted or "taper-pinned" to the tube. For example, a torque tube is generally used to couple and actuate the elevators, while a push-pull tube is used to couple and actuate double rudders.

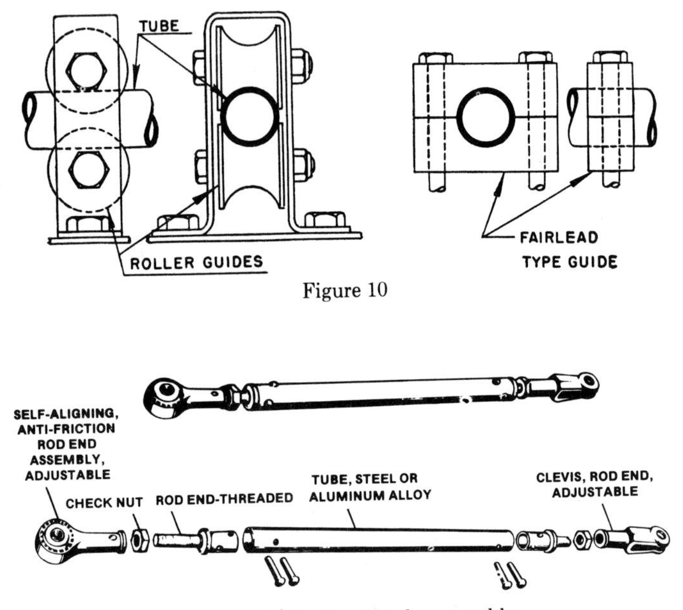

Figure 10

Figure 11. Push-pull tube assembly.

Push-pull tubes used to operate wing flaps or other control surfaces are generally spliced or attached to the actuating fittings by taper pins to facilitate service and repair. A push-pull tube carrying a heavy load is generally guided by Micarta bushings or rollers. This material will absorb a small amount of lubricant and serve as a bearing. The guides should always be carefully checked to be sure that they line up properly and do not draw the tube out of a straight line. For good results, the tube should fit the guides closely but freely.

Control cables are either flexible (7 × 7) or extraflexible (7 × 19). The latter is far more popular, except in sizes below ⅛ in. in diameter. 7 × 7 designates a cable composed of 7 strands, each composed of 7 wires; 7 × 19, a cable composed of 7 strands, each in turn composed of 19 wires. Control cables are generally made of stainless or galvanized steel, although some tinned steel cable still is used. The popular cable sizes are as follows:

1/16*	1/8	3/16	1/4	5/16
3/32*	5/32	7/32	9/32†	

* 7 × 7 only.
† 7 × 19 only.

Control cables are strung over AN210 pulleys and guided between pulleys by fair-leads when the span is long. Pulley brackets are equipped with cable guards to prevent the cable from dropping off the pulley when slackened.

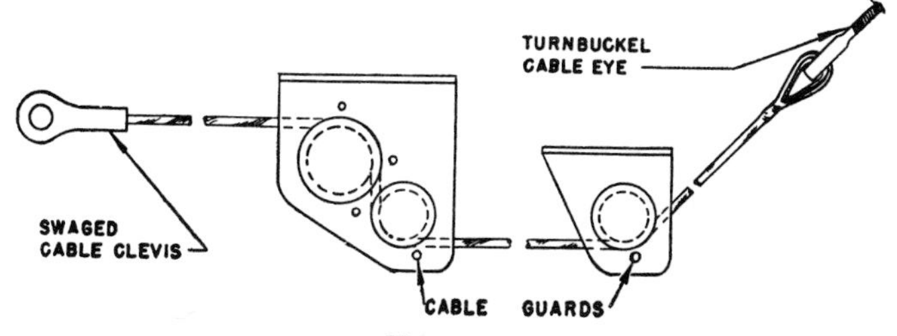

Figure 12.

Cable ends are either spliced around cable thimbles (AN100) or cable bushings (AN111), or the ends are swaged to cable terminals such as clevises, eyes, or threaded ends (AN666, 667, 668, and 669). The threaded ends are to fit in one end of a turnbuckle, the other ends to suit the installation.

When cables are being strung, it is usually necessary to remove the cable guards. After stringing, the cable should be checked to be sure that all guards are in place and properly safetied.

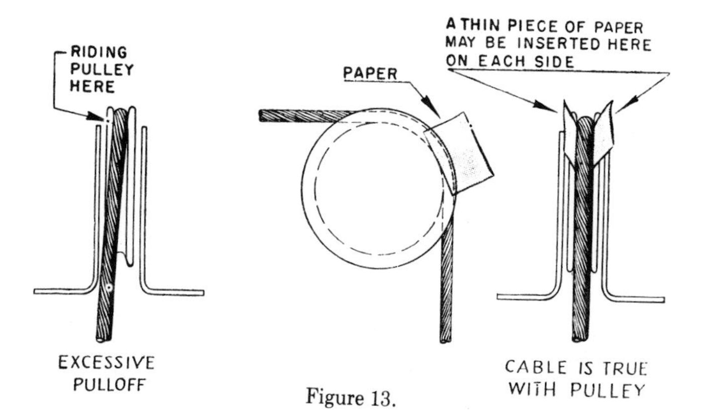

Figure 13.

A group of control cables are often strung very close together. It is therefore of utmost importance that each cable is strung through its proper pulley, that no cables are twisted, and that ends are in the right direction.

"Pull-off" is serious and must be avoided at all times. This term

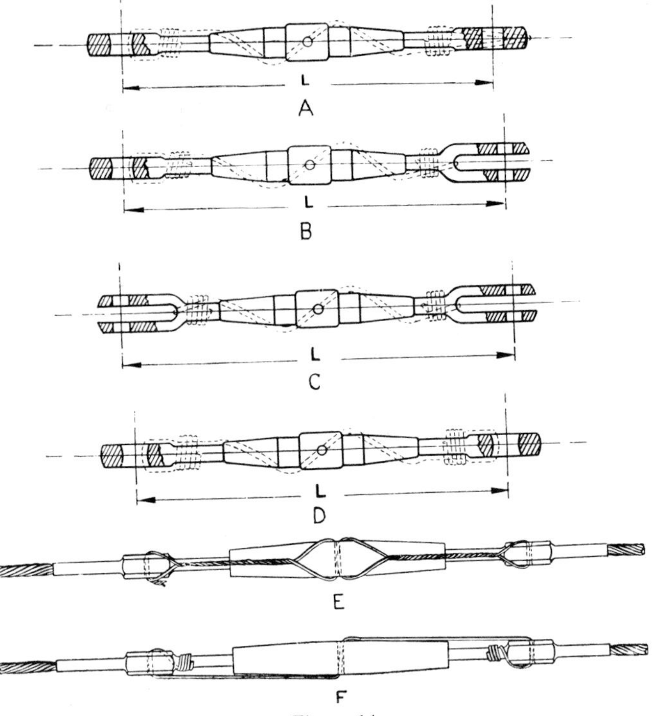

Figure 14.

denotes that the cable does not pull true with the pulley. Unless corrected, the cable will chafe the pulley sides and cause serious trouble. When a gauge to detect pull-off is not available, a thin piece of paper may be inserted between the cable and pulley side if there is no pull-off (Fig. 13).

To avoid pull-off, brackets should be set carefully by sighting through the pulley to the next pulley through which the cable runs, before the bolts holding the brackets are tightened. The system should be checked to see that cable terminals are properly secured and safetied, that cables have proper tension, and that turnbuckles are properly safetied (Fig. 14). The safety wire should tend to tighten the turnbuckle.

CHAPTER V
MATERIALS AND FABRICATING

ALUMINUM AND ALUMINUM ALLOYS

Aluminum is one of the most widely used metals in modern aircraft construction. It is vital to the aviation industry because of its high strength-to-weight ratio and its comparative ease of fabrication. The outstanding characteristic of aluminum is its light weight. Aluminum melts at the comparatively low temperature of 1,250° F. It is non-magnetic and is an excellent conductor.

Commercially pure aluminum is a white lustrous metal. Aluminum combined with various percentages of other metals forms alloys which are used in aircraft construction.

Aluminum alloys in which the principal alloying ingredients are either manganese, chromium, or magnesium and silicon show little attack in corrosive environments. Alloys in which substantial percentages of copper are used are more susceptible to corrosive action. The total percentage of alloying elements is seldom more than 6 or 7 percent in the wrought alloys.

Commercially pure aluminum has a tensile strength of about 13,000 p.s.i., but by rolling or other cold-working processes its strength may be approximately doubled. By alloying with other metals, or by using heat-treating processes, the tensile strength may be raised to as high as 65,000 p.s.i. or to within the strength range of structural steel.

Aluminum alloys, although strong, are easily worked because they are malleable and ductile. Most aluminum alloy sheet stock used in aircraft construction ranges from .016 to .096 inch in thickness; however, some of the larger aircraft use sheet stock which may be as thick as .356 inch.

The various types of aluminum may be divided into two general classes: (1) The casting alloys (those suitable for casting in sand, perma-

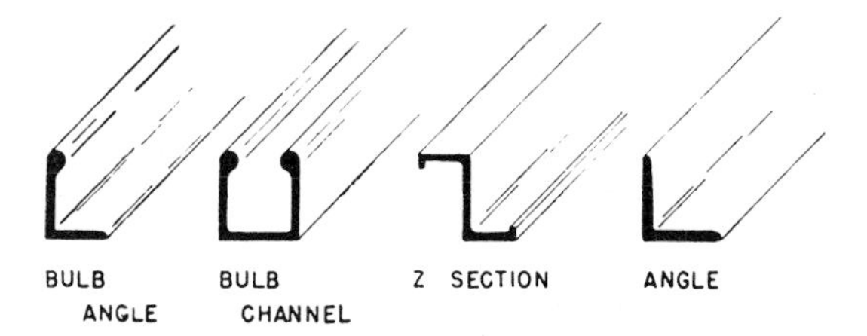

Fig. 1. EXTRUSIONS

nent mold, die castings), and (2) the wrought alloys (those which may be shaped by rolling, drawing, or forging). Of these two, the wrought alloys are the most widely used in aircraft construction, being used for stringers, bulkheads, skin, rivets, and extruded sections.

Aluminum Casting Alloys

Aluminum casting alloys are divided into two basic groups. In one, the physical properties of the alloys are determined by the alloying elements and cannot be changed after the metal is cast. In the other, the alloying elements make it possible to heat treat the casting to produce the desired physical properties.

The casting alloys are identified by a letter preceeding the alloy number. When a letter precedes a number, it indicates a slight variation in the composition of the original alloy. This variation in composition is simply to impart some desirable quality. In casting alloy 214, for example, the addition of zinc to improve its pouring qualities is indicated by the letter A in front of the number, thus creating the designation A214.

When castings have been heat treated, the heat treatment and the composition of the casting is indicated by the letter T, followed by an alloying number. An example of this is the sand casting alloy 355, which has several different compositions and tempers and is designated by 355-T6, 355-T51, or C355-T51.

Aluminum alloy castings are produced by one of three basic methods: (1) Sand mold, (2) permanent mold, or (3) die cast. In casting aluminum, different types of alloys are used for different types of castings. Sand castings and die castings require different types of alloys than those used in permanent molds.

Wrought Aluminum and Alloys

Wrought aluminum and wrought aluminum alloys are divided into two general classes, nonheat-treatable alloys and heat-treatable alloys.

Nonheat-treatable alloys are those in which the mechanical properties are determined by the amount of cold-work introduced after the final annealing operation. The mechanical properties obtained by cold working are destroyed by any subsequent heating and cannot be restored except by additional cold working, which is not always possible. The "full hard" temper is produced by the maximum amount of cold-work that is commercially practicable. Metal in the "as fabricated" condition is produced from the ingot without any subsequent controlled amount of cold working or thermal treatment. There is, consequently, a variable amount of strain hardening, depending upon the thickness of the section.

For heat-treatable aluminum alloys the mechanical properties are obtained by heat treating to a suitable temperature, holding at that temperature long enough to allow the alloying constituent to enter into solid solution, and then quenching to hold the constituent in solution. The metal is left in a supersaturated, unstable state and is then age hardened either by natural aging at room temperature or by artificial aging at some elevated temperature.

Typical mechanical properties of wrought aluminum alloys are shown in chart form on page 151.

Aluminum Designations

Wrought aluminum and wrought aluminum alloys are designated by a four-digit index system. The system is broken into three distinct groups: 1xxx group, 2xxx through 8xxx group, and 9xxx group (which is at present unused).

The first digit of a designation identifies the alloy type. The second digit indicates specific alloy modifications. Should the second number be zero, it would indicate no special control over individual impurities. Digit 1 through 9, however, when assigned consecutively as needed for the second number in this group, indicate the number of controls over individual impurities in the metal.

The last two digits of the 1xxx group are used to indicate the hundredths of 1 percent above the original 99 percent designated by the first digit. Thus, if the last two digits were 30, the alloy would contain 99 percent plus 0.30 percent of pure aluminum, or a total of 99.30 percent pure aluminum. Examples of alloys in this group are:

1100—99.00 percent pure aluminum with one control over individual impurities.
1130—99.30 percent pure aluminum with one control over individual impurities.
1275—99.75 percent pure aluminum with two controls over individual impurities.

In the 2xxx through 8xxx groups, the first digit indicates the major alloying element used in the formation of the alloy as follows:

2xxx—copper.
3xxx—manganese.
4xxx—silicon.
5xxx—magnesium.
6xxx—magnesium and silicon.
7xxx—zinc.
8xxx—other elements.

In the 2xxx through 8xxx alloy groups, the second digit in the alloy designation indicates alloy modifications. If the second digit is zero, it indicates the original alloy, while digits 1 through 9 indicate alloy modifications.

The last two of the four digits in the designation identify the different alloys in the group.

Where used, the temper designation follows the alloy designation and is separated from it by a dash; i.e., 7075-T6, 2024-T4, etc. The temper designation consists of a letter indicating the basic temper which may be more specifically defined by the addition of one or more digits. These designations are as follows:

—F As fabricated.
—O Annealed, recrystallized (wrought products only).
—H Strain hardened.
  —H1 (plus one or more digits) strain hardened only.
  —H2 (plus one or more digits) strain hardened and partially annealed.

—H3 (plus one or more digits) strain hardened and stabilized.
—W Solution heat treated, unstable temper.
—T Treated to produce stable tempers other than—F, —O, or —H.
—T2 Annealed (cast products only).
—T3 Solution heat treated and then cold worked.
—T4 Solution heat treated.
—T5 Artifically aged only.
—T6 Solution heat treated and then artificially aged.
—T7 Solution heat treated and then stabilized.
—T8 Solution heat treated, cold worked, and then artificially aged.
—T9 Solution heat treated, artificially aged, and then cold worked.
—T-10 Artificially aged and then cold worked.

In the wrought form, commercially pure aluminum is known as 1100. It has a high degree of resistance to corrosion and is easily formed into intricate shapes. It is relatively low in strength, however, and does not have the strength required for structural aircraft parts. Higher strengths are generally obtained by the process of alloying. The resulting alloys are less easily formed and, with some exceptions, have lower resistance to corrosion than 1100 aluminum.

Alloying is not the only method of increasing the strength of aluminum. Like other materials, aluminum becomes stronger and harder as it is rolled, formed, or otherwise cold-worked. Since the hardness depends on the amount of cold working done, 1100 and some wrought aluminum alloys are available in several strain-hardened tempers. The soft or annealed condition is designated O. If the material is strain hardened, it is said to be in the H condition.

The most widely used alloys in aircraft construction are hardened by heat treatment rather than by cold-work. These alloys are designated by a somewhat different set of symbols:—T4 and W indicate solution heat treated and quenched but not aged, and T6 indicates an alloy in the heat treated hardened condition.

Aluminum alloy sheets are marked with the specification number on approximaely every square foot of material. If for any reason this identification is not on the material, it is possible to separate the heat-treatable alloys from the nonheat-treatable alloys by immersing a sample of the material in a 10-percent solution of caustic soda (sodium hydroxide). The heat-treatable alloys will turn black due to the copper content, whereas the others will remain bright. In the case of clad material, the surface will remain bright, but there will be a dark area in the middle when viewed from the edge.

CLAD ALUMINUM

The terms "Alclad and Pureclad" (tradenames) are used to designate sheets that consist of an aluminum alloy core coated with a layer of pure aluminum to a depth of approximately 5½ percent on each side. The pure aluminum coating affords a dual protection for the core, preventing contact with any corrosive agents, and protecting the core electrolytically by preventing any attack caused by scratching or from other abrasions.

HEAT TREATMENT OF ALUMINUM ALLOYS

Heat treatment is a series of operations involving the heating and cooling of metals in the solid state. Its purpose is to change a mechanical property or combination of mechanical properties so that the metal will be more useful, serviceable, and safe for a definite purpose. By heat treating, a metal can be made harder, stronger, and more resistant to impact. Heat treating can also make a metal softer and more ductile. No one heat-treating operation can produce all of these characteristics. In fact, some properties are often improved at the expense of others. In being hardened, for example, a metal may become brittle.

The various heat-treating processes are similar in that they all involve the heating and cooling of metals. They differ, however, in the temperatures to which the metal is heated, the rate at which it is cooled, and, of course, in the final result.

Successful heat treating requires close control over all factors affecting the heating and cooling of metals. Such control is possible only when the proper equipment is available and the equipment is selected to fit the particular job. Thus, the furnace must be of the proper size and type and must be so controlled that temperatures are kept within the limits prescribed for each operation. Even the atmosphere within the furnace affects the condition of the part being heat treated. Further, the quenching equipment and the quenching medium must be selected to fit the metal and the heat-treating operation. Finally, there must be equipment for handling parts and materials, for cleaning metals, and for straightening parts.

Heat treating requires special techniques and equipment which are usually associated with manufacturers or large repair stations. Since these processes are usually beyond the scope of the field mechanic, the heat treatment of aluminum alloys will only be discussed briefly.

There are two types of heat treatments applicable to aluminum alloys. One is called solution heat treatment, and the other is known as precipitation heat treatment. Some alloys, such as 2017 and 2024, develop their full properties as a result of solution heat treatment followed by about 4 days of aging at room temperature. Other alloys, such as 2014 and 7075, require both heat treatments.

The alloys that require precipitation heat treatment (artificial aging) to develop their full strength also age to a limited extent at room temperature; the rate and amount of strengthening depends upon the alloy. Some reach their maximum natural or room-temperature aging strength in a few days, and are designated as—T4 or —T3 temper. Others continue to age appreciably over a long period of time. Because of this natural aging, the —W designation is specified only when the period of aging is indicated, for example 7075-W (½ hour). Thus, there is considerable difference in the mechanical and physical properties of freshly quenched (—W) material and material that is in the —T3 or —T4 temper.

The hardening of an aluminum alloy by heat treatment consists of four distinct steps:

1. Heating to a predetermined temperature.
2. Soaking at temperature for a specified length of time.
3. Rapidly quenching to a relatively low temperature.
4. Aging or precipitation hardening either spontaneously at room temperature, or as a result of a low-temperature thermal treatment.

The first three steps above are known as solution heat treatment, although it has become common practice to use the shorter term, "heat treatment." Room-temperature hardening is known as natural aging, while hardening done at moderate temperature is called artificial aging, or precipitation heat treatment.

Solution Heat Treatment

Temperature

The temperatures used for solution heat treating vary with different alloys and range from 825 degrees F. to 980 degrees F. (See chart on page 152). As a rule, they must be controlled within a very narrow range (plus or minus 10 degrees) to obtain specified properties. Heating is accomplished in either a fused salt bath or an air furnace. The soaking time varies, depending upon the alloy and thickness from 10 minutes for thin sheets to approximately 12 hours for heavy forgings. For the heavy sections, the nominal soaking time is approximately 1 hour for each inch of cross-sectional thickness. The soaking time is chosen so that it will be the minimum necessary to develop the required physical properties.

Quenching

After the soluble constituents are in solid solution, the material is quenched to prevent or retard immediate re-precipitation. Three distinct quenching methods are employed. The one to be used in any particular instance depends upon the part, the alloy, and the properties desired.

Cold Water Quenching

Parts produced from sheet, extrusions, tubing, small forgings, and similar type material are generally quenched in a cold water bath. The temperature of the water before quenching should not exceed 85 degrees F. Such a drastic quench ensures maximum resistance to corrosion. This is particularly important when working with such alloys as 2017, 2024, and 7075. This is the reason a drastic quench is preferred, even though a slower quench may produce the required mechanical properties.

Hot Water Quenching

Large forgings and heavy sections can be quenched in hot or boiling water. This type of quench minimizes distortion and alleviates cracking which may be produced by the unequal temperatures obtained during the quench. The use of a hot water quench is permitted with these parts because the temperature of the quench water does not critically affect the resistance to corrosion of the forging alloys. In addition, the resistance to corrosion of heavy sections is not as critical a factor as for thin sections.

Spray Quenching

High-velocity water sprays are useful for parts formed from clad sheet and for large sections of almost all alloys. This type of quench also minimizes distortion and alleviates quench cracking. However, many specifications forbid the use of spray quenching for bare 2017 and 2024 sheet materials because of the effect on their resistance to corrosion.

Lag Between Soaking and Quenching

The time interval between the removal of the material from the furnace and quenching is critical for some alloys and should be held to a minimum.

Re-heat Treatment

The treatment of material which has been previously heat treated is considered a re-heat treatment. The unclad heat-treatable alloys can be solution heat treated repeatedly without harmful effects.

The number of solution heat treatments allowed for clad sheet is limited due to increased diffusion of core and cladding with each re-heating. Existing specification allow one to three re-heat treatments of clad sheet depending upon cladding thickness.

Straightening After Solution Heat Treatment

Some warping occurs during solution heat treatment, producing kinks, buckles, waves, and twists. These imperfections are generally removed by straightening and flattening operations.

Where the straightening operations produce an appreciable increase in the tensile and yield strengths and a slight decrease in the percent of elongation, the material is designated—T3 temper. When the above values are not materially affected, the material is designated—T4 temper.

Precipitation Heat Treating

As previously stated, the aluminum alloys are in a comparatively soft state immediately after quenching from a solution heat-treating temperature. To obtain their maximum strengths, they must be either naturally aged or precipitation hardened.

Precipitation hardening produces a great increase in the strength and hardness of the material with corresponding decreases in the ductile properties. The process used to obtain the desired increase in strength is therefore known as aging, or precipitation hardening.

The aging practices used depend upon many properties other than strength. As a rule, the artificially aged alloys are slightly overaged to increase their resistance to corrosion. This is especially true with the artificially aged high-copper content alloys that are susceptible to intergranular corrosion when inadequately aged.

The heat-treatable aluminum alloys are subdivided into two classes, those that obtain their full strength at room temperature and those that require artificial aging.

The alloys that obtain their full strength after 4 or 5 days at room temperature are known as natural aging alloys. Precipitation from the supersaturated solid solution starts soon after quenching, with 90 percent of the maximum strength generally being obtained in 24 hours. Alloys 2017 and 2024 are natural aging alloys.

The alloys that require precipitation thermal treatment to develop their full strength are artificially aged alloys. However, these alloys also age a limited amount at room temperature, the rate and extent of the strengthening depending upon the alloys.

Many of the artificially aged alloys reach their maximam natural or room temperature aging strengths after a few days. These can be stocked for fabrication in the —T4 or —T3 temper. High-zinc content alloys such as 7075 continue to age appreciably over a long period of time, their mechanical property changes being sufficient to reduce their formability.

The advantage of—W temper formability can be utilized, however, in the same manner as with natural aging alloys; that is, by fabricating shortly after solution heat treatment, or retaining formability by the use of refrigeration.

Refrigeration retards the rate of natural aging. At 32 degrees F., the beginning of the aging process is delayed for several hours, while dry ice (-50 degrees F. to -100 degrees F.) retards aging for an extended period of time.

Precipitation Practices

The temperatures used for precipitation hardening depend upon the alloy and the properties desired, ranging from 250 degrees F. to 375 degrees F. They should be controlled within a very narrow range (plus or minus 5 degrees) to obtain best results. See chart on page 152.

The time at temperature is dependent upon the temperature used, the properties desired, and the alloy. It ranges from 8 to 96 hours. Increasing the aging temperature decreases the soaking period necessary for proper aging. However, a closer control of both time and temperature is necessary when using the higher temperatures.

After receiving the thermal precipitation treatment, the material should be air cooled to room temperature. Water quenching, while not necessary, produces no ill effects. Furnace cooling has a tendency to produce overaging.

ANNEALING OF ALUMINUM ALLOYS

The annealing procedure for aluminum alloys consists of heating the alloys to an elevated temperature, holding or soaking them at this temperature for a length of time depending upon the mass of the metal, and then cooling in still air. Annealing leaves the metal in the best condition for cold-working. However, when prolonged forming operations are involved, the metal will take on a condition known as "mechanical hardness" and will resist further working. It may be necessary to anneal a part several times during the forming process to avoid cracking. Aluminum alloys should not be used in the annealed state for parts or fittings.

Clad parts should be heated as quickly and carefully as possible, since long exposure to heat tends to cause some of the constituents of the core to diffuse into the cladding. This reduces the corrosion resistance of the cladding.

Heat-treatment of Rivets.—Rivets are made from various alloys and, of course, should be heat-treated before being used. In this condition, some alloyed rivets are too hard and will crack upon upsetting. This tendency may be controlled by retarding the age hardening of the rivets until the time of use. If they are placed in freezing boxes immediately after heating and quenching, their age hardening will be arrested and will continue when the rivets are removed and used in a structure. After their removal from the low temperature, they should be used within approximately 20 min before age hardening prevents riveting.

Fabricating

Forming.—The forming of aluminum is generally confined to sheet stock and employs such standard practice machines as drop hammers, presses, power brakes, and punch presses.

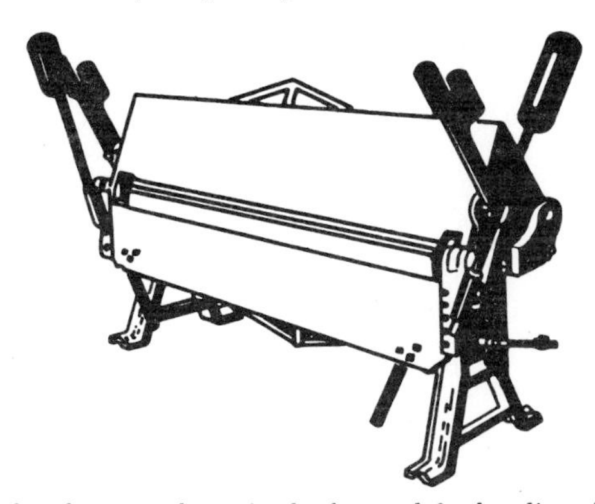

Fig. 2. A hand operated cornice brake used for bending sheet metal. Larger brakes are power operated.

Of most importance is the manner in which the dies or tools used on these machines are made and cared for during their operation, for the action of forming the metal must be done over exceedingly smooth and polished dies, using lubricants where necessary, such that the movement of the material over the tool surfaces will not scratch or smear the Alclad finish.

The simplest type of forming is the bending of a straight flange. Even this operation requires great care. First, in designing the part, consideration should be given to the sharpness of the bend, or bend radius, which depends on the temper of the material. For aluminum in the T condition this radius must not be less than 2½ times the thickness of the material; for O material the radius may be as little as the thickness of the material. It is practical to increase calculated minimum radii to conform to standard dies generally available in sizes varying by 1/32 in. Disregard for these minimum radii or the use of roughly finished tools may result in small flaws or cracks that will necessitate the rejection of the part.

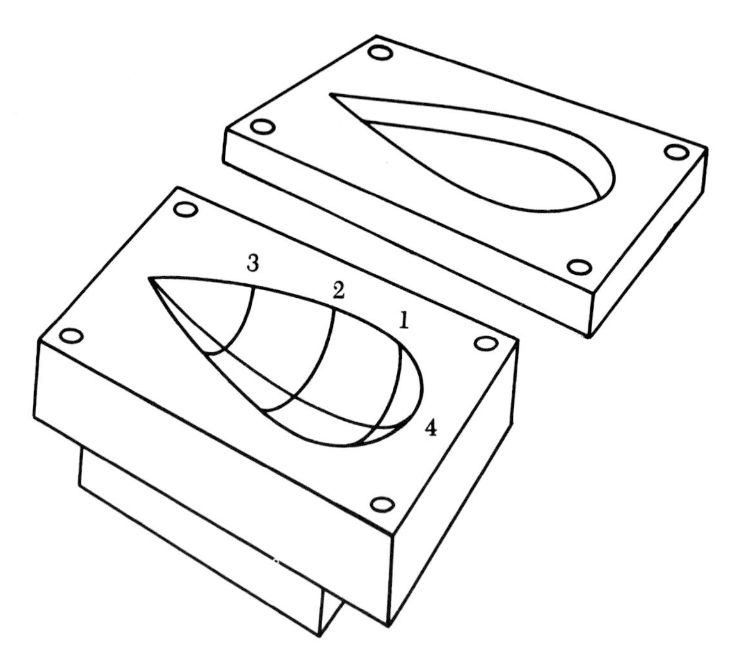

Fig. 3. A simple form block and holddown plate for hand forming. A speedier and better production is the use of the hydropress.

Bends of a more complicated design, as a sheet-metal rib having flanges around its contour, should be made over a form block shaped to fit the inside contour of the finished part. Bending the flanges over this die may be accomplished by "hand forming," a slow but practical method for experimental work.

A speedier and better production method is the use of the hydropress. The form block is laid on the base of the press with the material held in place by pin guides in the form block at locations where the pin holes will not affect the designed strength of the part. The upper platen of the press is covered with rubber from 4 to 6 in. thick. When it is lowered, a pressure of several tons is applied. The rubber forces the material edges down over the sides of the form block, either shrinking or stretching the material to fit its contour. This method, though used extensively, also has limitations such as the depth to which it may form

Fig. 4.—Double action hydraulic press, and deep drawn section of wing-tip tank. *(Courtesy of Lockheed Aircraft Corp., Burbank, Calif.)*

a part, and the amount of shrinking it will successfully perform before the material will wrinkle or fold over.

For such irregular shapes as cowlings and fairings where extreme forming is necessary, the drop hammer is applicable. Drop hammers use heavy dies and punches. The die, of Kirksite (a lead and zinc alloy) or steel, is generally cast to the shape of the part from a plaster pattern. After being finished by scraping or grinding, it may then be used to pour a lead punch. The drop hammer does not press the part into shape. The punch is raised by the upper platen and dropped into the die, forcing or ramming the metal to fit the shape of the die.

Parts that require deep forming, when fabricated on the drop hammer, often must be made progressively by the use of graduated dies forming the part in several stages. Another method uses the finish die in a progressive manner by the stacking of several layers of plywood rings between the punch and the die. This prevents the punch from bottoming until all the wood rings have been removed, one after each stroke. The falling of the punch on the rings serves two purposes; it stops the punch at a desired height and it holds the overlapping sheet metal in a flat plane, preventing wrinkling and thus producing a drawing action as each layer of plywood is removed.

FIG. 5.—Drop hammer showing sheet-metal part being formed to die contour by successive strokes or hits. (*Courtesy of Lockheed Aircraft Corp., Burbank, Calif.*)

Forming of irregular shapes may be speedily accomplished if the part is small and used in such quantity as to warrant making punch and dies for the punch press. The punch press is activated by a crankshaft, operated rapidly by engaging a clutch to a power-driven flywheel. Dies for punch press forming are more complicated and intricate than for other types of presses and generally perform a "drawing" operation. Such dies should be equipped with pressure pads that hold the material tightly against the face of the die while the punch draws the material in and down over the edges of the die.

Parts of a symmetrical concentric shape may often be best formed by spinning. Lathe chucks are formed to the inside contour of the part. Over these the metal is forced by pressure of a hard wood or polished steel spin tool. Spinning also is done on tubing that may need a flange turned outward on the end or an alteration in its diameter for special designs.

There are many special tools and applications of the foregoing machines to handle special forming jobs that arise from intricate designs. For instance, the power brake, though primarily used for bending straight flanges, has been adapted to use in forming dies for small parts and such special work as setting joggles in preformed parts and extrusions.

In the fabrication of aluminum where shrinking or stretching is severe, the metal is used in the O condition. After forming, the part is heat-treated to the T condition. Though the material used is in the O condition, the forming, if very severe, work-hardens the material. Often this work hardening is so pronounced as to hinder completion of the part; then it is necessary to anneal the material before forming is continued.

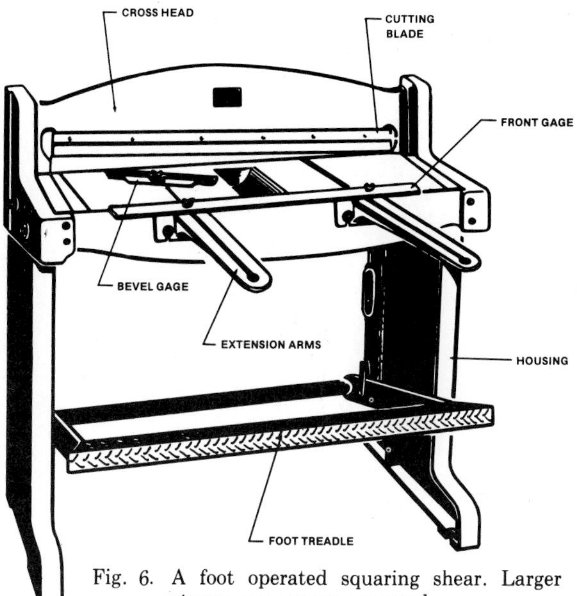

Fig. 6. A foot operated squaring shear. Larger shears are power operated.

Machining involves all forms of cutting, whether performed on sheet stock, castings, or extrusions, and involves such operations as shearing, sawing, routing, lathe and millwork, and such hand operations as drilling, tapping, and reaming.

Lathe and millwork and all turning operations performed on aluminum are governed by general machine-shop practices and differ from similar operations upon steels in that the turning is done at higher speeds taking lighter cuts. The cutting tools should be kept very sharp with more top and side rake than for steels. Cuts requiring a smooth finish generally are made under a flow of kerosene and lard oil lubricants. These oils are excellent for such hand cutting as drilling and tapping, but the more common soluble lubricants may be used equally as well for lighter work.

Cutting operations on sheet material are principally the blanking out of parts prior to forming or assembly. Small parts are blanked most quickly on a punch press while the larger patterns must be sheared, sawed, or routed.

Routing of irregularly shaped parts is one of the most extensively used practices, since it permits the cutting of several parts at one time by stacking several sheets. Using templates clamped to the material, a milling cutter of small diameter turning at high speed is guided around the template, routing a path equal to the diameter of the cutter. A small amount of lubricant supplied to the surface of the material makes a smoother cut and prolongs the life of the cutter.

Figure 7. Master router cutting several sheets of metal as it follows the large template clamped on the material. (*Courtesy of Lockheed Aircraft Corp., Burbank, Calif.*)

The router is adaptable to several types of machines. For the above-mentioned cutting, the router head is mounted on movable arms to follow the profile of the part. This machine is called the "master router." Another type is the "pin router" where the router head is stationary and the material is run against the cutter and guided by the template touching a pin directly below the cutter. Other stationary routers use jigs that nest small parts such as extruded sections, and on passing the cutter are trimmed, beveled, or tapered, and perform high-speed milling operations.

Welding

Flame welding of aluminum is generally confined to the non-heat-treatable alloys. Obviously, an alloy selected for high strength and used in the heat-treated condition should not have local portions partly annealed as a result of welding. Alloys 1100 & 3003, being nearly pure aluminum and practical for use in non-structural components such as fairings, air scoops, and cowlings, may be welded with practically no loss of their original strength. Of the heat-treatable alloys, 6053 may be welded quite satisfactorily. The welding action simulates that of heat-treatment, in that as the torch moves along there is a heating followed by an air quench which hardens this material approximately to its temper in the W condition.

The selection of the proper welding rod or wire is most important for good results. The non-heat-treatable alloy 3003 should ordinarily be welded with 3003 wire, while the heat-treatable alloy should ordinarily be welded with a silicon wire.

The production of a sound weld in aluminum necessitates the use of a flux to clean the material in the area of the weld and to remove the aluminum oxide present on the surface of the aluminum. The flux is generally a powder mixed with water and may be brushed on the surface to be welded, or the welding rod may be dipped in the solution prior to welding with satisfactory results. On completion of welds, this flux should be removed by washing, as the residual flux is corrosive to the material and its presence under a painted finish will lift the paint in a short while. If the welded area is readily accessible, hot water and brushing are sufficient. For inaccessible welds, the parts may be immersed in a solution of 10 per cent sulphuric acid for 30 min or 3 per cent sulphuric acid, heated to 150°F, for 10 min.

Resistance Welding.—Spot welding is practiced extensively on sheet aluminum of nearly all alloys. The principal requirements are good equipment with accurate current adjustment, timing devices to produce accurate and uniform spots, and material free of grease or dirt.

Cleaning is done with a fine abrasive cloth or light etching solution in the immediate area of the weld. Unless cleaned, the aluminum oxide on the surface of the material will offer high resistance and cause surface heating at the point of contact between the electrode and the aluminum, resulting in the electrodes either picking up a part of the aluminum surface or depositing some copper on the aluminum.

The spot-welding machine, equipped with copper electrodes for spot welding or rolls for seam welding, should be kept clean and smooth by frequent polishing. In polishing, extreme care should be used to keep a slight crown on the contact points. Rolls used in seam welding are electrically driven; timing devices form overlapping spots or seams. The timing may be adjusted to allow greater space between spots and thus will spot intermittently to any desired spacing.

Surface Protection

Pure aluminum is inherently resistant to corrosion. It is this quality that prompted the use of Alclad finishes to the alloyed materials. However, aircraft are often subject to adverse weather conditions and, especially on sea-going craft, it is important that as much protection as possible be afforded.

Anodic treatment is a means of oxydizing the surface of aluminum, providing a very thin and hard surface of aluminum oxide. This aluminum oxide provides a greater resistance to corrosion and an excellent base for paint primers to be applied later, if still greater protection is necessary.

Anodizing is electroplating in reverse. The aluminum part is made the anode in a chromic acid or sulphuric acid solution. By electrolysis, oxygen is deposited on the surface, forming aluminum oxide.

Various processes of anodizing are applicable depending on the finish desired. Alumiliting is an anodic process, practical for producing a finish that may include dye or pigment to provide color effects. It is often the final finish applied.

Chromodizing, a slightly less effective treatment that eliminates the expense of electrolysis, is accomplished by a simple chemical dip. Applied on such parts where requirements permit, it involves a 5-min immersion in a heated chromic acid bath.

MAGNESIUM AND MAGNESIUM ALLOYS

Magnesium, the world's lightest structural metal, is a silvery-white material weighing only two-thirds as much as aluminum. Magnesium does not possess sufficient strength in its pure state for structural uses, but when alloyed with zinc, aluminum, and manganese it produces an alloy having the highest strength-to-weight ratio of any of the commonly used metals.

Some of today's aircraft require in excess of one-half ton of this metal for use in hundreds of vital spots. Some wing panels are fabricated entirely from magnesium alloys, weigh 18 percent less than standard aluminum panels, and have flown hundreds of satisfactory hours. Among the aircraft parts that have been made from magnesium with a substantial savings in weight are nosewheel, doors, flap cover skin, aileron cover skin, oil tanks, floorings, fuselage parts, wingtips, engine nacelles, instrument panels, radio masts, hydraulic fluid tanks, oxygen bottle cases, ducts, and seats.

Magnesium alloys possess good casting characteristics. Their properties compare favorably with those of cast aluminum. In forging, hydraulic presses are ordinarily used, although, under certain conditions, forging can be accomplished in mechanical presses or with drop hammers.

Magnesium alloys are subject to such treatments as annealing, quenching, solution heat treatment, aging, and stabilizing. Sheet and plate magnesium are annealed at the rolling mill. The solution heat treatment is used to put as much of the alloying ingredients as possible into solid solution, which results in high tensile strength and maximum ductility. Aging is applied to castings following heat treatment where maximum hardness and yield strength are desired.

Magnesium embodies fire hazards of an unpredictable nature. When in large sections, its high thermal conductivity makes it difficult to ignite and prevents it from burning. It will not burn until the melting point is reached, which is 1,204 degrees F. However, magnesium dust and fine chips are ignited easily. Precautions must be taken to avoid this if possible. Should a fire occur, it can be extinguished with an extinguishing powder, such as powdered soapstone, or graphite powder. Water or any standard liquid or foam fire extinguishers cause magnesium to burn more rapidly and can cause explosions.

Magnesium alloys produced in the United States consist of magnesium alloyed with varying proportions of aluminum, manganese, and zinc. These alloys are designated by a letter of the alphabet, with the number 1 indicating high purity and maximum corrosion resistance.

HEAT TREATMENT OF MAGNESIUM ALLOYS

Magnesium alloy castings respond readily to heat treatment, and about 95 percent of the magnesium used in aircraft construction is in the cast form.

The heat treatment of magnesium alloy castings is similar to the heat treatment of aluminum alloys in that there are two types of heat treatment: (1) Solution heat treatment and (2) precipitation (aging) heat treatment. Magnesium, however, develops a negligible change in its properties when allowed to age naturally at room temperatures.

Solution Heat Treatment

Magnesium alloy castings are solution heat treated to improve tensile strength, ductility, and shock resistance. This heat-treatment condition is indicated by using the symbol —T4 following the alloy designation. Solution heat treatment plus artificial aging is designated —T6. Artificial aging is necessary to develop the full properties of the metal.

Solution heat-treatment temperatures for magnesium alloy castings range from 730 degrees F. to 780 degrees F., the exact range depending upon the type of alloy. The temperature range for each type of alloy is listed in Specification MIL-H-6857. The upper limit of each range listed in the specification is the maximum temperature to which the alloy may be heated without danger of melting the metal.

The soaking time ranges from 10 to 18 hours, the exact time depending upon the type of alloy as well as the thickness of the part. Soaking periods longer than 18 hours may be necessary for castings over 2 inches in thickness. Magnesium alloys must never be heated in a salt bath as this may result in an explosion.

A serious potential fire hazard exists in the heat treatment of magnesium alloys. If through oversight or malfunctioning of equipment, the maximum temperatures are exceeded, the casting may ignite and burn freely. For this reason, the furnace used should be equipped with a safety cutoff that will turn off the power to the heating elements and blowers if the regular control equipment malfunctions or fails.

Some magnesium alloys require a protective atmosphere of sulphur dioxide gas during solution heat treatment. This aids in preventing the start of a fire even if the temperature limits are slightly exceeded.

Air-quenching is used after solution heat treatment of magnesium alloys since there appears to be no advantage in liquid cooling.

Precipitation Heat Treatment

After solution treatment, magnesium alloys may be given an aging treatment to increase hardness and yield strength. Generally, the aging treatments are used merely to relieve stress and stabilize the alloys in order to prevent dimensional changes later, especially during or after machining. Both yield strength and hardness are improved somewhat by this treatment at the expense of a slight amount of ductility. The corrosion resistance is also improved, making it closer to the "as cast" alloy.

Precipitation heat-treatment temperatures are considerably lower than solution heat-treatment temperatures and range from 325 degrees F. to 500 degrees F. Soaking time ranges from 4 to 18 hours.

TITANIUM AND TITANIUM ALLOYS

In aircraft construction and repair, titanium is used for fuselage skins, engine shrouds, firewalls, longerons, frames, fittings, air ducts, and fasteners. Titanium is used for making compressor disks, spacer rings, compressor blades and vanes, through bolts, turbine housings and liners, and miscellaneous hardware for turbine engines.

Titanium falls between aluminum and stainless steel in terms of elasticity, density, and elevated temperature strength. It has a melting point of from 2,730 degrees F. to 3,155 degrees, low thermal conductivity, and a low coefficient of expansion. It is light, strong, and resistant to stress-corrosion cracking. Titanium is approximately 60 percent heavier than aluminum and about 50 percent lighter than stainless steel.

Because of the high melting point of titanium, high-temperature properties are disappointing. The ultimate yield strength of titanium drops rapidly above 800 degrees F. The absorption of oxygen and nitrogen from the air at temperatures above 1,000 degrees F. makes the metal so brittle on long exposure that it soon becomes worthless. However, titanium does have some merit for short-time exposure up to 3,000 degrees F. where strength is not important. Aircraft firewalls demand this requirement.

Titanium is nonmagnetic and has an electrical resistance comparable to that of stainless steel. Some of the base alloys of titanium are quite hard. Heat treating and alloying do not develop the hardness of titanium to the high levels of some of the heat-treated alloys of steel. It was only recently that a heat-treatable titanium alloy was developed. Prior to the development of this alloy, heating and rolling was the only method of forming that could be accomplished. However, it is possible to form the new alloy in the soft condition and heat treat it for hardness.

Iron, molybdenum, and chromium are used to stabilize titanium and produce alloys that will quench harden and age harden. The addition of these metals also adds ductility. The fatigue resistance of titanium is greater than that of aluminum or steel.

TITANIUM DESIGNATIONS

The A-B-C classification of titanium alloys was established to provide a convenient and simple means of describing all titanium alloys. Titanium and titanium alloys possess three basic types of crystals: A (alpha), B (beta), and C (combined alpha and beta). Their characteristics are:

A (alpha)—All-around performance; good weldability; tough and strong both cold and hot, and resistant to oxidation.

B (beta)—Bendability; excellent bend ductility; strong both cold and hot, but vulnerable to contamination.

C (combined alpha and beta for compromise performances)—Strong when cold and warm, but weak when hot; good bendability; moderate contamination resistance; excellent forgeability.

Titanium is manufactured for commercial use in two basic compositions; commercially pure titanium and alloyed titanium. A-55 is an example of a commercially pure titanium. It has a yield strength of 55,000 to 80,000 p.s.i. and is a general-purpose grade for moderate to severe forming. It is sometimes used for nonstructural aircraft parts and for all types of corrosion resistant applications, such as tubing.

Type A-70 titanium is closely related to type A-55 but has a yield strength of 70,000 to 95,000 p.s.i. It is used where higher strength is required, and it is specified for many moderately stressed aircraft parts. For many corrosion applications, it is used interchangeably with type A-55. Both type A-55 and type A-70 are weldable.

One of the widely used titanium-base alloys is designated as C-110M. It is used for primary structural members and aircraft skin, has 110,000 p.s.i. minimum yield strength, and contains 8 percent manganese.

Type A-110AT is a titanium alloy which contains 5 percent aluminum and 2.5 percent tin. It also has a high minimum yield strength at elevated temperatures with the excellent welding characteristics inherent in alpha-type titanium alloys.

CORROSION CHARACTERISTICS

The corrosion resistance of titanium deserves special mention. The resistance of the metal to corrosion is caused by the formation of a protective surface film of stable oxide or chemi-absorbed oxygen. Film is often produced by the presence of oxygen and oxidizing agents.

Corrosion of titanium is uniform. There is little evidence of pitting or other serious forms of localized attack. Normally, it is not subject to stress corrosion, corrosion fatigue, intergranular corrosion, or galvanic corrosion. Its corrosion resistance is equal or superior to 18-8 stainless steel.

HEAT TREATMENT OF TITANIUM

Titanium is heat treated for the following purposes:

1. Relief of stresses set up during cold forming or machining.
2. Annealing after hot working or cold working, or to provide maximum ductility for subsequent cold working.
3. Thermal hardening to improve strength.

Stress Relieving

Stress relieving is generally used to remove stress concentrations resulting from forming of titanium sheet. It is performed at temperatures ranging from 650 degrees F. to 1,000 degrees F. The time at temperature varies from a few minutes for a very thin sheet to an hour or more for heavier sections. A typical stress-relieving treatment is 900 degrees F. for 30 minutes, followed by an air cool.

The discoloration or scale which forms on the surface of the metal during stress relieving is easily removed by pickling in acid solutions. The recommended solution contains 10 to 20 percent nitric acid and 1 to 3 percent hydrofluoric acid. The solution should be at room temperature or slightly above.

Full Annealing

The annealing of titanium alloys provides toughness, ductility at room temperature, dimensional and structural stability at elevated temperatures, and improved machinability.

The full anneal is usually called for as preparation for further working. It is performed at 1,200 degrees F. to 1,650 degrees F. The time at temperature varies from 16 minutes to several hours, depending on the thickness of the material and the amount of cold work to be performed. The usual treatment for the commonly used alloys is 1,300 degrees F. for 1 hour, followed by an air cool. A full anneal generally results in sufficient scale formation to require the use of caustic descaling, such as sodium hydride salt bath.

Thermal Hardening

Unalloyed titanium cannot be heat treated, but the alloys commonly used in aircraft construction can be strengthened by thermal treatment, usually at some sacrifice in ductility. For best results, a water quench from 1,450 degrees F., followed by re-heating to 900 degrees F. for 8 hours is recommended.

Casehardening

The chemical activity of titanium and its rapid absorption of oxygen, nitrogen, and carbon at relatively low temperatures make casehardening advantageous for special applications. Nitriding, carburizing, or carbonitriding can be used to produce a water-resistant case of 0.0001 to 0.0002 inch in depth.

STEEL

Source

Steel is a metal resulting from the purification of iron and the reduction of its carbon content. Iron is obtained from ore extracted from the earth in various degrees of impurity and purified and refined by smelting to oxidize the carbon and other impurities. Steel is iron containing less than 2 per cent of carbon. Other elements present in steel are manganese, important to the increase of toughness, and silicon, acting as a gas eliminator to prevent blowholes. Also present are sulphur and phosphorus in very small amounts.

Classification

The alloying of steels by adding other metals, singly or in combination, results in alloys of varied uses and properties. These many alloys are classified by an S.A.E. numbering system to identify them in drawings, specifications, etc.

The principal alloy combinations are classified by basic numerals as follows:

Alloy	Number	Alloy	Number
Carbon steels	1	Chromium steels	5
Nickel steels	2	Chromium-vanadium steels	6
Nickel-chromium steels	3	**Tungsten** steels	7
Molybdenum steels	4	Silicomanganese steels	9

The use of these assigned numerals in coding steel specifications is as follows. The class of the alloy is indicated first in the code; for example in the code 4130, the class is a molybdenum steel indicated by the number 4. The second numeral, 1, indicates the percentage of this predominating alloying element, 1 per cent chromium. The last two numerals, 30, indicate the average carbon content in hundredths of 1 per cent. or .30 per cent.

Plain carbon steels, not being an alloy steel, may be indicated as follows:

Steel 1020: The first numeral, 1, indicates carbon steel. The second numeral, 0, indicates no alloy. The last two numerals, 20, indicate .20 per cent carbon range.

Note: Carbon range is considered to vary 5 points above or below its coded number. Thus, in the example, the carbon range is .15 to .25 per cent carbon.

The first two numerals, indicating the type and percentage of the predominating element, may necessarily require three numerals instead of two. For example, after a first numeral of 7 (tungsten steel), it may contain 13 per cent tungsten, indicated 713; if the carbon range is .50 per cent, it becomes 71350.

Identification

Supplementing the S.A.E. classification of steels, the Air Forces have adapted a standard method of readily identifying material in stock by painting color combinations on the surface. Parallel stripes are painted at the ends and in the middle of bar and sheet stock. These bands are combinations of broad and narrow stripes of different colors. The broad stripes (4 to 5 in. wide) indicate the first two or first three digits of the code. The narrow stripes (2 in. wide) indicate the last two digits of the code.

By referring to the following table: to identify 1020 steel, select the color red, from the broad stripe column, indicating 10, and from the narrow stripe column, the color yellow indicating 20. These colors are painted across the material separated by a 1-in. space. In other cases, it may require two colors to identify one pair of numerals. In this case, each color will take one-half of the 5-in. stripe if indicating the first two numerals, or one-half of the 2-in. stripe if indicating the second pair of numerals.

Figure 9 shows such a sample in the coding of 4130 steel.

COLOR SCHEME FOR MARKING STEEL

Broad stripes, 4 to 5 in. wide		Narrow stripes, 2 in. wide	
Red	10	Red and black	00
Red and white	12	Red	10
Red and yellow	13	Red and green	12
Yellow	23	Red and white	15
Yellow and green	25	Yellow	20
Green	31	Yellow and white	25
Blue	32	Black	30
Brown	33	Black and white	35
Black	34	Green	40
Black and white	41	Green and white	45
Red and black	46	Black and green	46
Khaki	51	Blue	50
Red and blue	53	Brown	60
White	61	Brown and white	65
Red and brown	72	Khaki	95
Blue and yellow	76		
Blue and brown	92		
Brown and white	512		
Brown and yellow	521		
Purple	713		
Purple and yellow	716		

The addition of an orange stripe indicates annealed stock. The addition of a gray stripe indicates heat-treated stock. This color scheme does not apply to corrosion-resistant steels, since they are not classified by S.A.E. numbers.

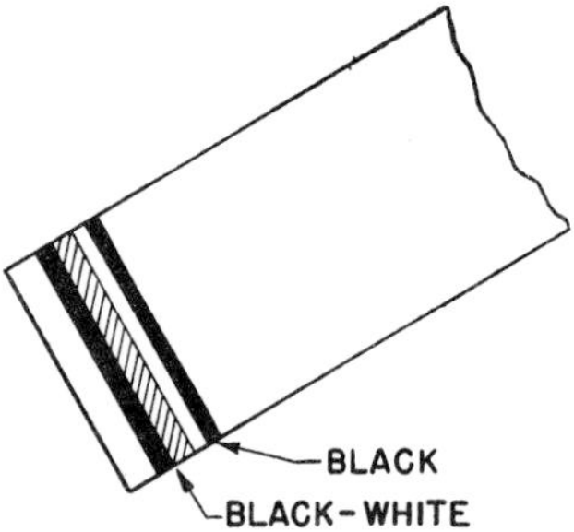

BLACK
BLACK-WHITE

FIG. 9.

PROPERTIES

The properties and uses of commonly used steels in aircraft are as follows:

1010–1025, low-carbon steels are soft and ductile steels used for low stressed parts where cold working is required Fittings made of these steels may be readily machined and welded but are not adaptable to heat-treatment, except casehardening.

1050–1095, high-carbon steels are heat-treatable after forming or machining and are used for parts requiring high shear strength and wearing surfaces such as drills, taps, and similar hand tools or springs, provided heat-treatment is done after forming.

2330–2350, nickel steels are heat-treatable after fabrication and are excellent for parts requiring high stress and wear, such as bolts, clevises, turnbuckles, and small fittings.

4130–4150, molybdenum steels, high-strength steels used in sheet form and tubing, are employed extensively for fuselage and landing-gear structures where heat-treatment is impractical owing to structural size or shape. It has the excellent quality of retaining most of its strength after being welded. Fittings requiring higher strength may be fabricated and heat-treated to high strengths.

Corrosion- and Heat-resistant Steels

General.—Stainless steels are used where corrosion resistance and heat resistance are important factors in the design of the part. There are many varieties of stainless steels and variations in the basic analysis and, as yet, S.A.E. classifications have not been assigned to all types. The A.I.S.I. (American Iron and Steel Institute) has however, assigned numbers to several of the commonly used alloys.

Basically, stainless steels contain 18 per cent of chromium and 8 per cent of nickel and are commonly referred to as 18-8 steels. These steels can be either flame- or spot-welded, are drawn and formed easily, are non-heat-treatable, but are available in various degrees of hardness from annealed to full hard condition, the result of strain hardening by cold rolling or cold drawing in their manufacture.

The commonly used alloys have A.I.S.I. numbers as follows, and also a common name designation.

A.I.S.I. Number	Common Designation
302	18-8
303	18-8 free machining
321	Stabilized titanium
347	Stabilized columbium
None	Inconel

Identification.—Since these steels do not have S.A.E. classifications for all alloys, the steel color scheme for identifying is not applicable, but they are rubber-stamped with A.I.S.I. numbers. Example: 302 ½H is 18-8 corrosion-resistant steel, ½ hard. Inconel must be stamped Inconel followed by 1A or 1H. The properties of commonly used corrosion-resistant steels are as follows:

302 alloy is most commonly used in the machining of sheet-metal parts. It is available in the annealed condition, designated 1A, or various degrees of hardness, ¼H, ½H, ¾H, or 1H (fully hardened).

303 alloy contains selenium or zirconium to reduce the work-hardening characteristics of 18-8 steels and to provide easier machining.

321 alloy contains titanium to increase its corrosion resistance when subjected to welding or other extreme temperatures.

347 alloy contains columbium, resulting in characteristics similar to 321 steels.

Inconel is another corrosion-resistant steel, of high nickel content, having the advantage of easier formability with slight work-hardening characteristics. It does not tend to corrode at welded joints, nor is it susceptible to inner granular corrosion when subjected to high temperatures; therefore it is best adaptable to exhaust collectors, etc.

Fabrication

Forming of steel is accomplished by any of the methods and machines described under Aluminum, the essential difference being the workability of the material used. Steel is a stronger material and requires harder dies and more powerful machines, according to the type of steel being formed.

The stronger steels, such as 4130, are more difficult to work and, where possible, are limited to simple forming; assembly is completed by welding or bolting. Forming is easier in the case of low-carbon steels, stainless steels, or Inconel; these being more ductile steels and well suited to drawing operations.

The bend radii of steels are subject to minima, particularly the high-strength alloys. The grain in high-strength steels is almost always apparent to the eye; whenever possible they should be bent across the grain.

Cutting of sheet steel is done principally by shearing and sawing or by punch press dies. Routing is not applicable except at slow speeds. For the trimming of irregular parts, there are several types of small machines and power-driven hand tools performing shearing or nibbling operations. The nibbler is used extensively for trimming the ends of tubing where intersecting tubes necessitate a developed curve. The nibbler uses a small (1/8 to 1/4 in. in diameter) punch and die, which punch circular blanks from the part as it is fed into the machine, cutting a path equal to the diameter of the punch. The rough scalloped edge left on the part requires only a small amount of handwork to finish smooth.

Machining of steel parts and fittings for aircraft must follow the best machine-shop practices and requires a knowledge of proper setups, cutting requirements of various metals, cutting speeds, and sharpening of tools. Machining operations are varied, owing to the wide range of hardness of different steels. On page 46 are tables of cutting speeds for different materials, which will aid in setting the turning speeds according to the diameter of the part.

The automatic lathe is the most highly productive type in aircraft shops but requires tedious setup. The turret lathe is most common; while not so fast as the automatic lathe, it requires considerably less setup time. The engine lathe requires least setup but is so slow that it is used principally for tool work and small lots.

For parts requiring exceedingly close tolerances and smooth finish, grinding is the common practice; for close fits, as required by portions of the hydraulic systems, honing and lapping produce the desired finish.

Heat-treatment

Heat-treatment of steels involves the heating and cooling of the metal; the rate of this heating and cooling determines the crystalline structure of the material. Almost all metals have a critical temperature at which the grain structure changes. Steel at its critical temperature changes structure when carbon and iron are said to form a solid solution. Common forms of heat-treatment are hardening, tempering, annealing, normalizing, and casehardening.

Hardening.—To obtain maximum hardness, the steel should be heated to a temperature in excess of its critical temperature to ensure a complete change of state (upper critical temperature). Exceeding this temperature by 25 to 50°F is necessary to ensure thorough heating of the inside of the piece. Rapid quenching from this temperature will result in maximum hardness.

The required temperature of heating varies for different alloys, from 1500 to 1650°F; for best results, parts should be put in the oven when its heat is approximately 1000°F and gradually increased to its hardening temperature. See page 153 for hardening temperatures and maximum tensile strength obtained for various common alloys. Quenching is accomplished in water, air, oil, or salt brine. Oil results in the slowest cooling rate but is sometimes necessary for alloys that require higher heating temperatures to prevent cracking and excess warping from the strain of uneven cooling.

Accurate methods of measuring and controlling temperatures of the oven should be available for best results; however, where necessary, the change in color can be used to determine temperatures. The color chart, page 154, gives temperatures corresponding to colors assumed by the metal as it is heated.

Tempering (Drawing).—Metal that has been hardened by rapid cooling from a point above its critical temperature is often harder than necessary, too brittle for most purposes, and under internal strains. In order to relieve the strains and reduce brittleness, the metal is usually tempered after being hardened. This is accomplished by heating the hardened steel to a temperature below the critical range (400 to 1200°F). The degree of strength and the hardness remaining depend on the temperature of heating, the less this temperature the more hardness remaining. The table on page 153 shows the drawing temperatures for common alloys and the corresponding tensile strength. Tempering, like hardening, requires accurate methods of heating and measuring temperatures for good results, but the temperatures may also be approximated by noting the change of color as the temperature rises. These colors appear in the well-cleaned or polished surface accomplished by buffing or grinding to remove the oxide film formed during hardening. Refer to the color chart for drawing temperatures, page 154.

Annealing of steels, the reverse of hardening, is performed when necessary to reduce or remove hardness and to increase their ductility. Heating to the critical temperature removes any hardness caused by previous heat-treatments or working strains. From this critical temperature, slow cooling will restore them to a state of minimum hardness. This is accomplished simply by allowing them to cool with the furnace. Annealing is often done to allow severe forming of the material; on completion of the part, it may be hardened and tempered again to a desired strength.

Normalizing is a special case of annealing for the purpose of removing strains in fabricated parts induced by machining, bending, or welding. It is accomplished by heating to a point above its critical temperature and allowing to cool in still air, avoiding drafts that would cause uneven cooling and set up strains again in the part.

Casehardening, a special treatment for iron base alloys, produces a hard surface but leaves the core tough and resilient. This may be accomplished by carburizing, nitriding, or cyaniding.

Carburizing is done by heating the metal while it is in contact with a solid, liquid, or gas, rich in carbon. Several hours of this treatment is required for the surface of the material to absorb carbon and thus become a high-carbon steel.

Nitriding is a similar process applied to special steels. The heated material is held in contact with anhydrous ammonia. Iron nitrides, formed in the surface of the metal, produce a greater hardness than carburizing but to lesser depth and only in certain special steels.

Cyaniding is a rapid method of casehardening by immersing the heated steel in a molten bath of cyanide or applying powdered cyanide to the surface of the material.

Surface Protection

Surface protection of steels is generally accomplished by painting, by plating with a corrosion-resistant metal, or by plating followed by painting. In any case, an essential prerequisite is the proper cleaning of the material.

Cleaning of steel parts prior to painting necessitates the removal of all grease or oil formerly applied for corrosion resistance during shipment or for a manufacturing lubricant. Degreasing may be done most simply by washing in a solvent such as carbon tetrachloride followed by thorough drying; if facilities are available, the parts may be vapor cleaned. In this process the parts are passed through tanks containing vaporized solvent, which condenses on the parts, dissolves, and carries away the grease or oil.

Pickling is a chemical cleaning process, involving an acid or electrolytic dip, applied to nearly all metals for cleaning. It varies in its formula and method for various metals and cannot be considered a standard process. This treatment, if not followed immediately by washing and painting, must be washed in an alkaline solution to neutralize the acid.

For stainless steels, the pickling solution is a strong acid, which will loosen particles of steel, scale of welding, or materials imbedded from forming dies. Stainless steel fabricated parts generally require pickling, particularly after welding, when scale must be loosened and removed by brushing.

Passivate.—Stainless steel used in manifolds or in areas subject to heat may be further protected from corrosion by an oxidizing finish resulting from passivating, which is a chemical dip producing a very thin and invisible oxide surface.

Sandblasting is a fast method of removing scale and dirt by an air blast carrying sand or finely divided steel particles, frequently applied on steel parts previously welded. This method when employed on stainless steels should not contain steel particles as some may be imbedded in the material and later corrode.

Plating.—Probably the most extensively used method of preventing corrosion on small fittings is cadmium plating, an electrolytic process depositing a layer of cadmium on the surface of the part. This, of course, is impractical if the parts are too large or if the part is of a laminated nature where the acid solution may enter crevices from which it cannot be washed out. Chromium plating is also used for corrosion protection but is generally applied where use can be made of its hard surface for wearing qualities or, if desired, for its appearance.

Painting may be applied directly to the cleaned steel and be the only corrosion protection or it may be applied after any of the other protecting processes. A paint primer is first applied, generally zinc chromate P-27, followed by colored lacquers if desired for appearance.

Greasing.—Wearing surfaces of machined parts, such as bearings and bushings, should be protected from painting and any of the acid treatments by masking or plugging. Later they will require an application of grease or oil to protect the surface during stocking.

HARDNESS TESTING

Hardness testing of both raw materials and finished parts is necessary to determine whether the materials and the heat-treatment have met the strength values used in design. By one of three common methods, Rockwell, Brinell, or Shore scleroscope, a close approximation can be made of a material's tensile strength, which is closely related to its hardness. Tables on page 155 show these approximate relations for carbon and low-alloy steels and can be used in a general manner for other alloys. The hardness of steels is indicated by a number derived from tests on materials of a known tensile strength.

Fig. 10.—Rockwell hardness tester. (*Courtesy of Wilson Mechanical Instrument Co., New York, N.Y.*)

Rockwell testing for hardness is accomplished by a device that applies a known load to a penetrator on the surface of the material being tested. A minor load is first applied seating the penetrator in the material; penetration is then effected by the application of a major load. A dial mounted on the instrument will measure this penetration, and, as the pointer comes to a rest, the major load is released; the pointer will then return to a position indicating the depth still penetrated by the minor load. The greater the difference between these two readings, the less the hardness number and the softer the material.

The penetrator may be a 1/16-dia. steel ball with a 100-kg weight for the major load, or a diamond-cone penetrator with a 150-kg weight. The former is used for the softer materials and indicated on a red *B* scale about the circumference of the dial; readings for the diamond penetrator are taken from a black *C* scale. The minor load is 10 kg with either penetrator.

Brinell testing for hardness is done with a small hydraulic press applying a known load (3,000 kg for steels) to a steel ball that makes a spherical impression in the surface of the material being tested. The area of the impression is measured by a microscope with micrometer eyepiece. The load in kilograms divided by the area of the impression in square millimeters gives the Brinell number.

Shore scleroscope testing for hardness consists of dropping a diamond-tipped hammer on the material from a definite height and measuring the rebound. Several tests are generally made and an average of all readings is taken as the Shore number.

X RAY

In this inspection an actual x-ray photograph of the part is studied for flaws. This examination is generally applied to castings where gas pockets, slag inclusions, cracks, or other faults of casting might be present. X-ray inspection is generally required only on designated parts, either highly stressed or vital to the airplane's operation. Whether a 10 per cent inspection is needed to determine the "run" of the castings in a special lot, or 25, or 100 per cent, is specified by the structures and process engineers.

MAGNAFLUX

Magnaflux inspection is applicable to ferromagnetic materials, such as iron and steel, and is a dependable method of detecting cracks and flaws in or near the surface of fabricated parts. Highly stressed steel parts are generally magnafluxed to detect cracks resulting from heat-treatment or machining strains, or welding, which may leave cracks resulting from the heating and cooling of the metal or from improper welding procedure.

The parts to be magnafluxed are first magnetized, setting up flux lines around the part. If the part is cracked, opposite poles are formed at the break in the material. Finely divided ferromagnetic particles sprayed over the part will tend to gather at this break and form a pattern outlining its boundaries. An operator inspecting parts by this method should be skilled and experienced in judging which defects are cause for rejection. Irregularly shaped parts having sharp corners or machined recesses tend to cause confusing indications and should be carefully analyzed before judgment is passed.

Magnafluxing equipment consists of electrical apparatus for magnetizing parts, a tank holding a supply of kerosene or special oil in which the ferromagnetic particles are suspended, and a spray system to apply this solution to the parts being tested. After inspection, the parts are demagnetized by passing them through an alternating field (demagnetizing coils) and then rinsed clean.

CHAPTER VI

AIRCRAFT DRAWINGS

Interpreting a drawing and visualizing the appearance of a part or assembly necessitate an understanding of drafting practices and of the principles of orthographic projection, which establish the methods of illustrating and dimensioning a part.

The following notes are intended to aid the production mechanic only in determining from the print how the part is made, not how the print was drawn.

Although the various manufacturers' drafting systems will differ in detail, there is central agreement in the broad arrangement of drawings. This arrangement serves as an index for the quick location of the specific information required from the print being studied.

A line drawing of the part itself makes up the greater portion of the print; supplemented by dimensions and notes, it completely describes the part. In the lower right-hand corner is a block, referred to as the "legend" or "title block," containing reference material pertaining to the part. The information here given should be the first portion of the print to be analyzed.

The legend or title block will contain such information as:

Name of the print—identifies the part.

Number of the print—indexes this part for filing reference.

Model—of the airplane, or of the airplane unit if an aircraft accessory.

Signatures of draftsmen, engineer, checker, and project engineer.

Material—notes its form, dimensions, and pertinent specifications.

Finish—notes any painting, anodizing, plating required, and determines the final condition and appearance of the part. (This information may be indicated by code, which varies with each manufacturer.)

Scale—indicated as full, half, etc., giving the proportion of the drawing to the actual part.

Tolerance—designates the degree of precision necessary in fabricating the part and guides the inspector in determining the conformance to the print.

Immediately outside the legend are found such notes as are necessary to label the views or to elaborate any information not sufficiently detailed in the legend.

Change block is the space allotted to recording design changes. Located directly over the legend or at the top right of the print, it records the date of each change in the drawing and is important when supplying replacement parts for existing aircraft.

The foregoing facts may be applied to interpret the illustrations of the drafting practices that follow.

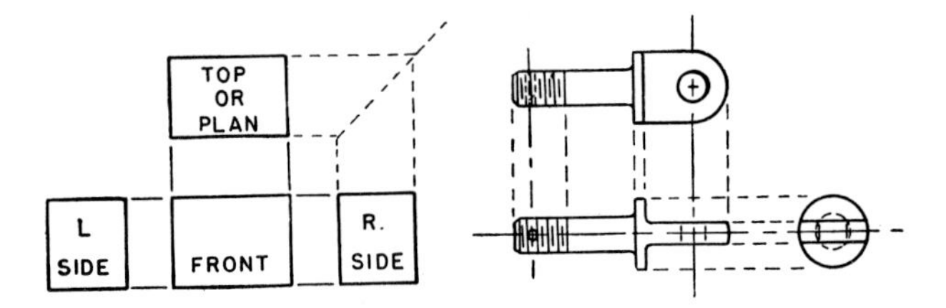

FIG. 1.

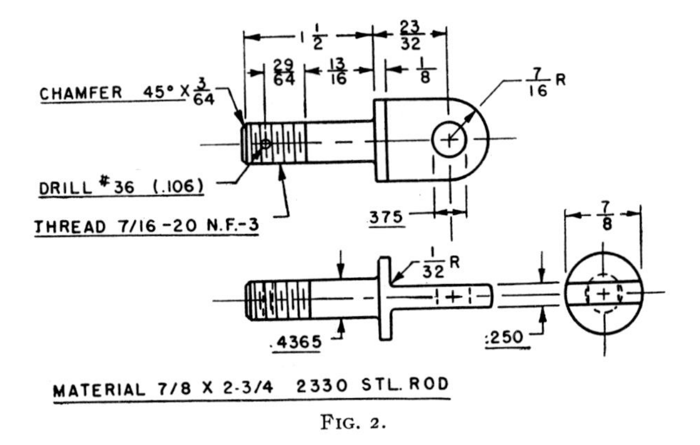

FIG. 2.

Orthographic projection is almost universally accepted as the most accurate method to describe fully a part or assembly. Figure 1 illustrates the arrangement of views. The front, side, and top views are arranged with the top (or plan) view directly over the front, and the side views directly to the side and in line with the front view.

It will be noted, in the part drawn at the right, that only those views needed to describe the part are shown. The left view is omitted since it adds no information not given in the right side view. Generally, three views, or even two or one, may be sufficient. The illustration alone does not, however, fully describe the part if dimensional and material information is lacking. Drawings should be complete and give every requirement for making the part.

The completed drawing (Fig. 2) follows such standard drafting room practices as follows:

Index of Lines.—Lines are drawn of different boldness and composition to identify the assortment of boundary lines, center lines, hidden lines, etc., needed to represent the part.

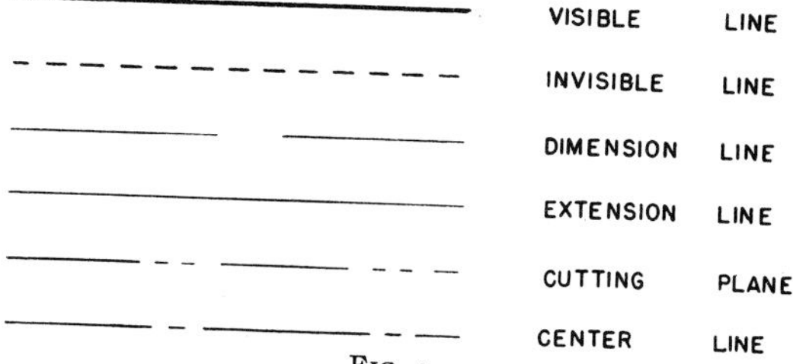

FIG. 3.

Dimensioning.—It will first be noted that dimensions show the size or length of the part, not the size of the drawing. This rule should

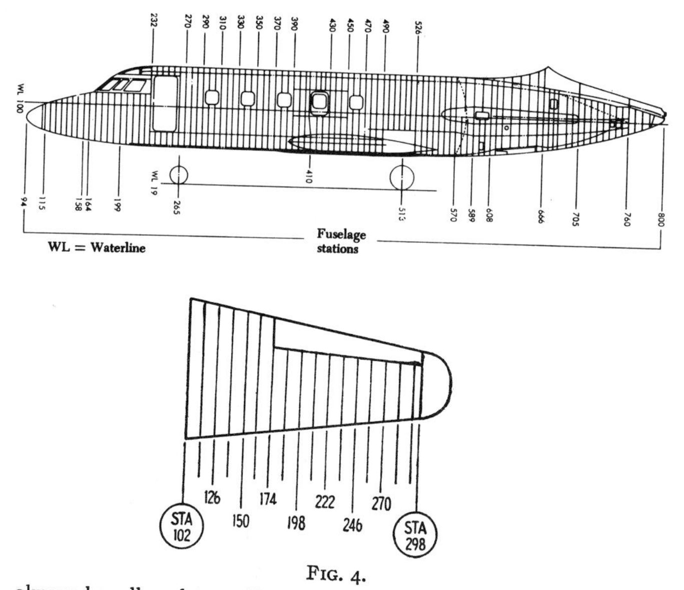

FIG. 4.

always be adhered to. Although the drawing, if to full scale, would coincide with the part, it is poor practice to scale a drawing to take a dimension.

The dimension line will always be drawn parallel to the dimension indicated and be bounded by extension lines at right angles. The dimension is printed in the break of the dimension line and always reads horizontally regardless of the direction of the dimension line. Dimensions on aircraft drawings are always given in inches, even when the full airplane length is given.

Stations are established at definite points along the fuselage and outward along the wing to aid in locating parts and to make it unnecessary to draw long and overlapping dimension lines. These stations are not placed at random but at such important structural members as ribs and bulkheads. Although in a few drafting systems station designations have no dimensional significance, fuselage stations usually are measurements in inches from the nose of the aircraft aft (from the fire wall, if a single-engine aircraft); wing stations are from the center line of the airplane measuring outboard.

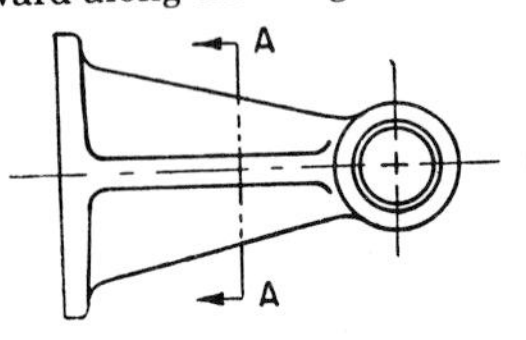

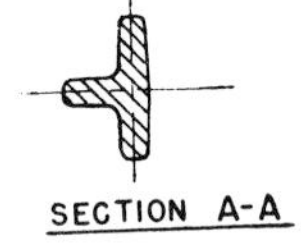

FIG. 5.

Station lines will be found on other drawings than that of the completed airplane; for example, the drawing of an aileron will show station lines at rib locations. The station line dimensions will however be dimensioned from the center line of the airplane, as on the wing, and will coincide with the same stations along the wing.

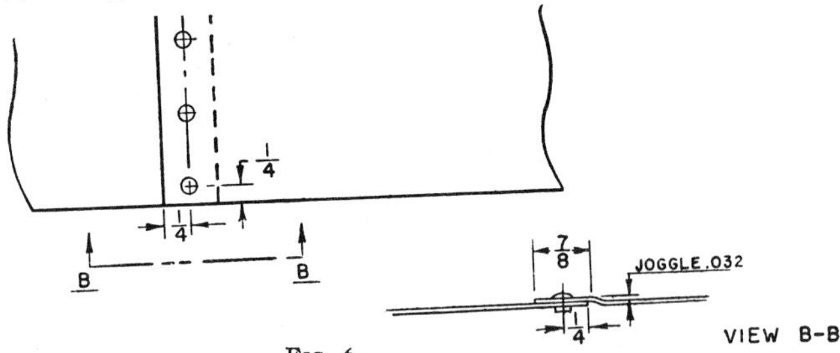

FIG. 6.

Sectional views (Fig. 5) are drawn to show the cross-sectional profile of a part. The location from which this view is taken is indicated by a cutting plane line which suggests that the part is cut in two at this line. The view of the exposed surface is rotated to an end view. The arrows at the ends of the cutting plane point toward the surface to be shown.

The sectional view is crosshatched according to the drafting room practices of the manufacturer to indicate the material. However, to eliminate confusing lines, sheet metal and such thin sections are not crosshatched.

The illustration of view *B-B* (Fig. 6) is another application of the cutting plane line. Not actually a cross section of a part, it is used to eliminate the necessity of drawing a rotated view of the complete part.

Here only the approximate portion represented by the length of the cutting plane is given in a rotated view to show the detail of the splice in this particular part. The letters used in the illustration do not identify the type of view but alphabetically identify and locate the sectional view on the principal view.

WORKING DRAWINGS

Working drawings must give such information as size of the object and all of its parts, its shape and that of all of its parts, specifications as to the material to be used, how the material is to be finished, how the parts are to be assembled, and any other information essential to making and assembling the particular object.

Working drawings may be divided into three classes: (1) Detail drawings, (2) assembly drawings, and (3) installation drawings.

Detail Drawing A detail drawing is a description of a single part, given in such a manner as to describe by lines, notes, and symbols the specifications as to size, shape, material, and methods of manufacture that are to be used in making the part. Detail drawings are usually rather simple: and, when single parts are small, several detail drawings may be shown on the same sheet or print.

Assembly Drawing An assembly drawing is a description of an object made up of two or more parts. It describes the object by giving, in a general way, the size and shape. Its primary purpose is to show the relationship of the various parts. An assembly drawing is usually more complex than a detail drawing, and is often accompanied by detail drawings of various parts.

Installation Drawing An installation drawing is one which includes all necessary information for a part or an assembly of parts in the final position in the aircraft. It shows the dimensions necessary for the location of specific parts with relation to the other parts and reference dimensions that are helpful in later work in the shop.

PICTORIAL DRAWINGS

A pictorial drawing is similar to a photograph. It shows an object as it appears to the eye, but it is not satisfactory for showing complex forms and shapes. Pictorial drawings are useful in showing the general appearance of an object and are used extensively with orthographic projection drawings. Pictorial drawings are used in maintenance and overhaul manuals.

CHAPTER VII

LOFT AND TEMPLATES

LOFTING

Layout in the *loft* consists of accurately laying out the shapes and contours of the aircraft. Layout in the *engineering department* consists of accurately laying out the various structure and installations to match the shape of the aircraft or the functional situation. Layout in the *template department* consists of accurately laying out the patterns with which the sheet-metal parts are fabricated (cut to size, formed, and drilled.)

The loft and template methods are briefly described here to give an insight into their relationship with shop practice.

The loft determines the contour or shape of the airplane after basic dimensions are furnished by the engineering department. For instance, the engineers decide, because of customer requirements, that the airplane will be of a certain length, height, and width. Also that it must have certain other dimensions along its length. The loft lays these dimensions out on the loft floor and then establishes a set of reference lines known as stations, water lines, and buttock lines as follows:

A horizontal line, through the center of the aircraft, in its flying position, is called "water line o." Parallel to and above and below this line, at intervals of 3 or 6 in., other water lines are drawn known as WL 1, WL 2, etc., or WL—1, WL—2, etc., if they lie below the original line.

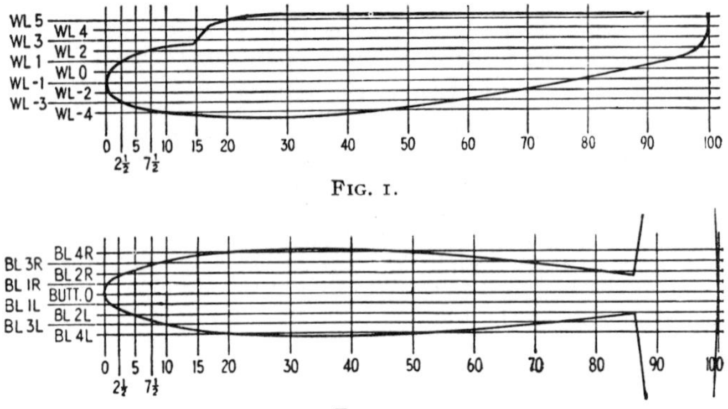

FIG. 1.

FIG. 2.

Next beginning at the extreme left, which represents the nose station lines are usually placed at 2½, 5, 7½, 10, 15, 20, 30, 40, 50, 60, 70, 80, 90 and 100 per cent of the aircraft's length. We now have the aircraft divided vertically and horizontally on a side view as shown in Fig. 1.

Now imagine that you are looking down on top of the aircraft, seeing a plan view. This also is divided into stations the same as the side view. A center line lengthwise through the aircraft is established, called "buttock line o." Lines to the right and left are located at 3 to 6-in. intervals to the right and left of buttock line o, called "Butt 1L" or "Butt 2R" etc. (Fig. 2).

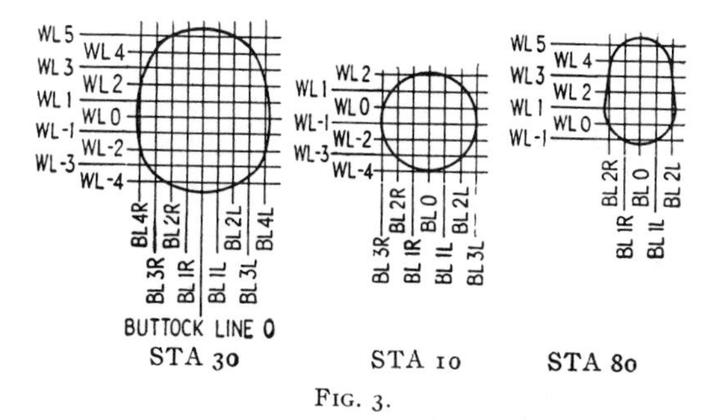

FIG. 3.

In a front view at the various stations these reference lines appear as shown in FIG. 3.

By running faired (smooth curved) lines through these stations and intersections, a clean airplane is conceived. All position dimensions within the airplane will now use these as reference and limits.

After the airplane has been lofted, to avoid confusion the layout engineers establish a new set of stations, located this time in increments of inches from the nose of the ship or some other fixed point in its vicinity. Thus, parts located in relation to the nose might be at Sta. 96 or Sta. 212, meaning 96 or 212 in. from the reference point at the nose of the aircraft.

Parts or points located to the right or the left are in reference to buttock line o, now called the "center line" (℄) of the ship; and those located vertically are located in inches from water line o.

The contour or shape of each individual rib or airfoil in the wing must be lofted from similar basic dimensions.

The *chord line* is used as the reference line and is described as a straight line drawn from the foremost point of the airfoil to the most rearward point.

The *airfoil* is the contour, above and below the chord line.

Engineering information is given that establishes the length of the chord line and gives dimensions, up and down, from the chord line at percentages of the chord line length. After dividing the chord line in

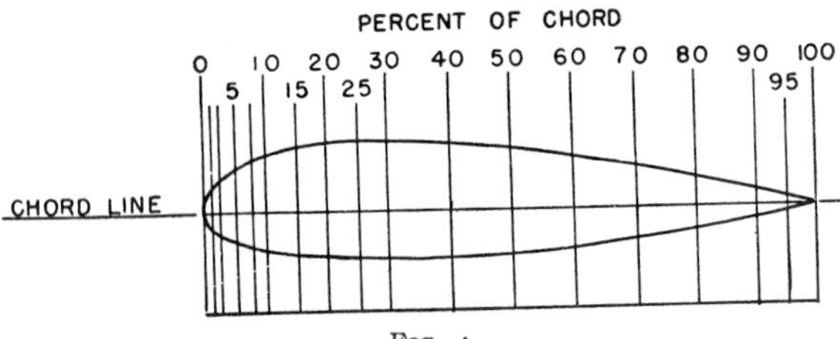

FIG. 4.

these percentages and measuring up and down from the chord line to locate the contour points, a faired line may be drawn to determine the shape (Fig. 4).

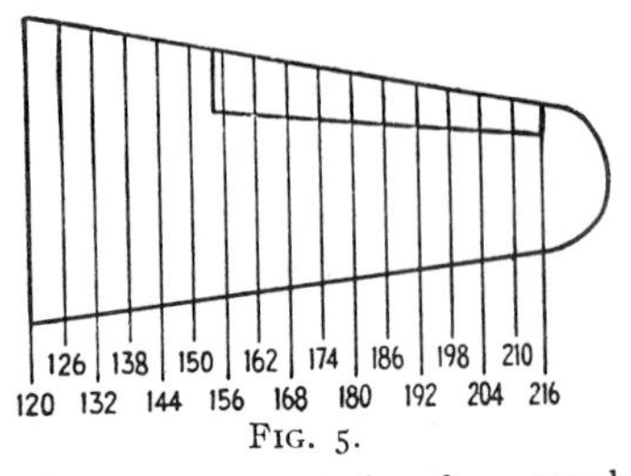

FIG. 5.

The wing and other lifting or control surfaces are also laid out by stations, which are generally designated at rib locations and are measured from the ℄ of the aircraft occasionally from the root of the wing (Fig. 5).

TEMPLATES

Template layout, as mentioned before, consists of making patterns for the fabrication of parts. They are made on metal, terneplate (lead-coated sheet iron), or galvanized iron. The lead coating allows the maker to scribe a fine line, which leaves the bright lead to contrast sharply with the painted surface, generally green, red, or black.

The patterns are developed on the metal; then the template is cut out and accurately filed to split the scribed line. Master templates, used to make other templates, are generally painted red to distinguish them from others.

Although there are many developments involving complicated layouts, a good share of the developments are straight lines and may be found by the following procedure:

Flat pattern layout by bend allowance (BA) is most universally accepted. By the use of the bend allowance chart and the following equations, any straight bend may be accurately developed.

A part with flanges, for example, is dimensioned to the mold lines as is the channel in Fig. 6 *A*.

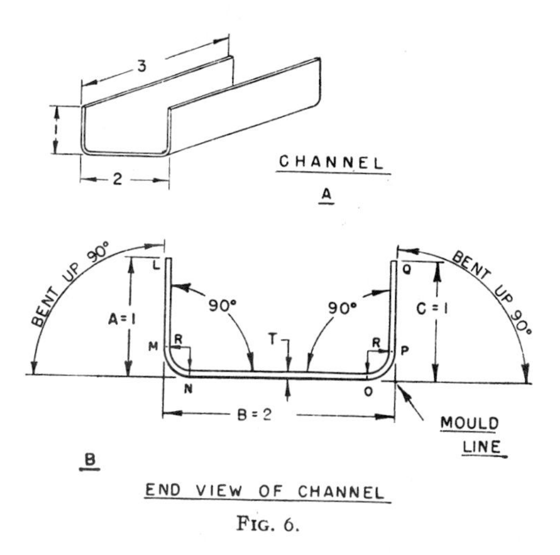

FIG. 6.

The mold lines (ML) indicate the point to which the dimension is really given and, you might say, the boundaries of the part (Fig. 6*B*).

The length of the channel will, of course, remain unchanged from flat pattern to finished part. On the other hand, the material necessary to make the channel web and the two flanges will not be the sum of their dimensions in the finished part, as it might at first seem, but rather, this "developed length" will be less than that sum, owing to the short cuts taken around the curved corners.

One might figure that each corner is a quarter of a circle and calculate its length in that manner, but practice has developed the *empirical formula* for bend allowance, which is much easier. The procedure for its use is

1. Imagine that the length around the cross section of the channel is divided into sections according to the letters L, M, N, O, P, and Q (Fig. 6*B*). Length LM = 1 in. − R (radius of bend) − T (thickness of metal). If $R = 3/16$ and $T = 1/16$ (.064), then

$$LM = 1 \text{ in.} - (3/16 \text{ in.} + 1/16 \text{ in.}) = 3/4 \text{ in.}$$

Lay this length out along the metal as shown in Fig. 7 and draw the bend line (BL).

2. Next, for the length MN, called the "bend allowance," refer to the bend allowance chart, page 104, for 1-deg bend allowance and multiply it by the number of degrees of bend (90 deg). The chart gives .00377 for $3/16R$ and $.064T$; when multiplied by 90, we get .3393 or $11/32$ (closest fraction) to the next bend line.

3. The web of the channel has two bends to be subtracted. Therefore it is 2 in. − $2(T + R)$ or 2 in. − ½ in. = 1½ in. from N to O.

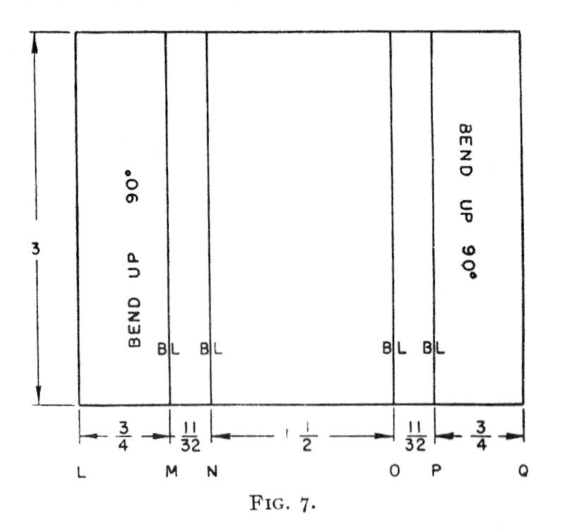

FIG. 7.

Repeat the first two operations in reverse order to complete the pattern.

A computation of this kind for aircraft is generally done in decimals for greater accuracy. Therefore, using decimals, the lengths would be

$A = 1.0000$	$B = 2.0000$	$C = 1.0000$
$T + R = .2515$	$2TR = .5030$	$T + R = .2515$
$LM = .7485$	$NO = 1.4970$	$PQ = .7485$

$$MN = OP = \text{bend allowance} = 90^\circ \cdot .00377 = .3393$$

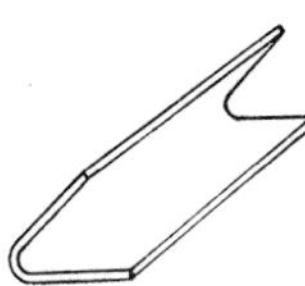

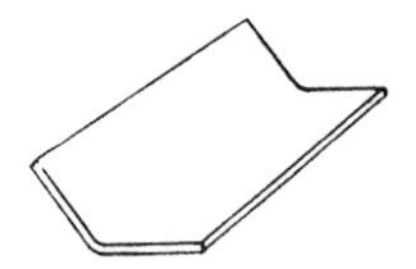

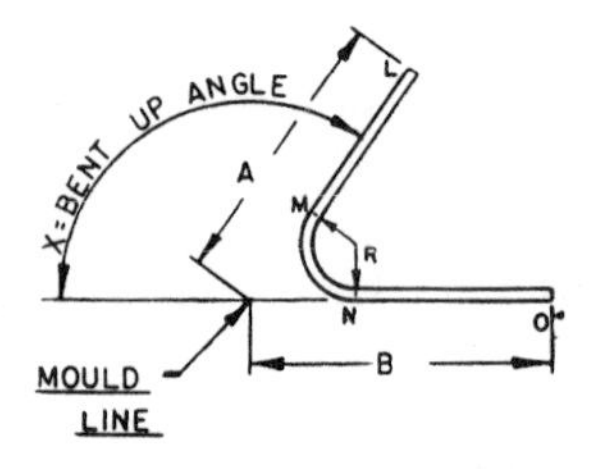

CLOSED BEVEL

OPEN BEVEL

FIG. 8.

Therefore, the developed length (sum $LM + MN + NO + OP + PQ$) equals 3.6726. Referring back, this is less than the sum of the dimensioned sides in Fig. 6*B*.

Formed angles of open or closed bevels may be developed in the same manner except that the distance from the corner mold line to the bend line is found by the formula

$$(T + R) \times (\text{the tangent of half the "bent-up" angle})$$

(Fig. 8).

BEND ALLOWANCE DERIVATION

The coefficients for bend allowance are computed from the bend allowance empirical formula. This formula, as the name implies, was developed by experimentation after a mathematical formula had been found, as shown below:

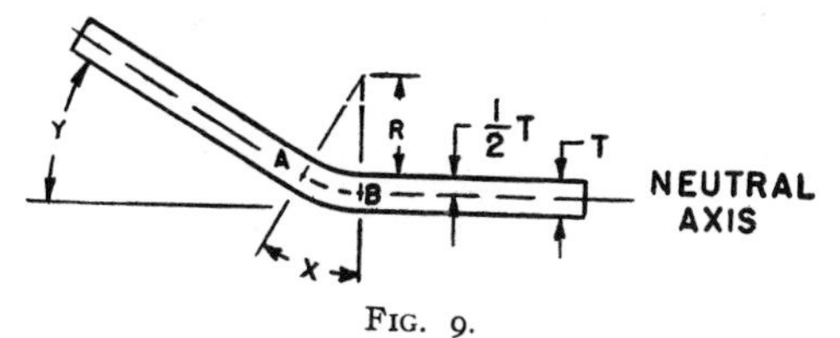

FIG. 9.

Bend allowance = BA along neutral axis

$$BA = 2\pi \cdot \left(R + \frac{T}{2}\right) \cdot \frac{X}{360}$$

$$BA = (2\pi R + \pi T) \cdot \frac{X}{360}$$

$$BA = \frac{(2\pi R + \pi T) \cdot X}{360}$$

$$BA = \frac{(6.2832R)}{(360)} + \frac{(3.1416T)}{(360)} \cdot X$$

$$BA = (.01743R + .00872T) \cdot X$$

The neutral axis was found, by experimentation, to lie slightly inside the $1/2T$, therefore the formula $(.01743R + .0078T)$ gives the bend allowance for 1-deg bend. The table (*page 104*) gives the 1-deg bend allowances for certain given radii and thicknesses. These are multiplied by the number of degrees bend to arrive at the total bend allowance.

Values are for 1-deg bend. Values derived from

$$\text{B.A.} = (.0078T + .01743R)N$$

Example: $T = .032$, $R = \frac{1}{8}$, angle $= 90°$:

$$\text{B.A.} = .00243 \times 90 = .2187.$$

SHEET-METAL BEND ALLOWANCE CHART FOR 1-DEG ANGLE

R \ T	.020 .022	.023 .025	.028 .029	.031 .032	.038 .040	.050 .051	.063 .064	.081	.091 .094	.125 .129
1/32	00072	00073	00076	00079	00086	00094	00104	00117	00125	00154
1/16	00126	00128	00131	00135	00140	00149	00159	00172	00180	00209
3/32	00180	00183	00185	00188	00195	00203	00213	00226	00234	00263
1/8	00235	00237	00240	00243	00249	00258	00268	00281	00289	00317
5/32	00290	00292	00294	00297	00304	00312	00322	00335	00343	00372
3/16	00344	00346	00349	00352	00358	00367	00377	00390	00398	00426
7/32	00398	00401	00403	00406	00412	00421	00431	00444	00452	00481
1/4	00454	00455	00458	00461	00467	00476	00486	00500	00507	00535
9/32	00507	00510	00512	00515	00521	00530	00540	00553	00561	00590
5/16	00562	00564	00567	00570	00576	00584	00595	00608	00616	00644
11/32	00616	00619	00620	00624	00630	00639	00649	00662	00670	00699
3/8	00671	00673	00675	00679	00685	00693	00704	00717	00725	00753
13/32	00725	00728	00730	00733	00739	00748	00758	00771	00779	00808
7/16	00780	00782	00784	00767	00794	00802	00812	00826	00834	00862
15/32	00834	00836	00839	00842	00848	00857	00867	00853	00888	00917
1/2	00889	00891	00893	00896	00903	00911	00921	00935	00943	00971
17/32	00943	00945	00948	00951	00957	00966	00976	00989	00997	01025
9/16	00998	01000	01002	01005	01012	01020	01030	01044	01051	01080
19/32	01051	01054	01055	01058	01065	01073	01083	01098	01105	01133
5/8	01107	01109	01111	01114	01121	01129	01139	01152	01160	01189
21/32	01161	01163	01166	01170	01175	01183	01193	01207	01214	01245
11/16	01216	01218	01220	01223	01230	01238	01248	01261	01269	01298
23/32	01269	01272	01273	01276	01283	01291	01301	01316	01322	01351
3/4	01324	01327	01329	01332	01338	01347	01357	01370	01378	01407
25/32	01378	01381	01383	01386	01392	01401	01411	01425	01432	01461
13/16	01433	01436	01438	01441	01447	01456	01466	01479	01487	01516
27/32	01487	01490	01491	01494	01501	01509	01519	01534	01540	01569
7/8	01542	01545	01548	01550	01556	01565	01575	01588	01596	01625
29/32	01596	01600	01601	01604	01610	01619	01629	01643	01650	01679
15/16	01651	01654	01655	01657	01665	01674	01684	01697	01705	01734
31/32	01705	01708	01709	01712	01718	01727	01737	01752	01758	01787
1	01760	01763	01765	01768	01774	01783	01793	01806	01814	01834

Minimum allowance below line ———————— for hard dural.

Minimum allowance below line - - - - - - - - - - for soft dural and steel.

CHAPTER VIII
STANDARD PARTS

STANDARD PARTS IDENTIFICATION

Since the manufacture of aircraft requires a large number of miscellaneous small fasteners and other items usually called "hardware," some degree of standarization is required. These standards have been derived by the various military organizations and described in detail in a set of specifications with applicable identification codes. These military standards have been universally adopted by the civil aircraft industry.

The derivation of a uniform "standard" is of necessity an evolutionary process. Originally, each of the military services derived their own standards. The old Army Air Corps set up AC (Air Corps) standards whereas the Navy used NAF (Naval Aircraft Factory) standards. In time, these were consolidated into AN (Air Force-Navy) standards and NAS (National Aerospace Standards). Still later these were consolidated into MS (Military Standard) designations.

At present, the three most common standards are:

° AN, Army-Navy
° MS, Military Standard
° NAS (National Aerospace Standards)

The aircraft mechanic however, will also occasionally be confronted with the following standard parts on older aircraft:

° AC (Air Corps)
° NAF (Naval Aircraft Factory)

Each of these standard parts is identified by its specification number and various dash numbers and letters to fully describe its name, size and material.

Additional information on AN, MS, NAS as well as AMS and AND specifications and schedule of prices for specification sheets can be obtained from: National Standards Associatiom, 1321 Fourteenth St. N.W., Washington, D.C. 20005.

Most airframe manufacturers have need for special small parts and use their own series of numbers and specifications. However, they use the universal "Standard Parts" wherever practicable.

Since the purpose of this "Standard Aircraft Handbook" is to provide the mechanic with a handy reference, only the most common "Standard Parts" are mentioned here with sufficient information to identify them.

STANDARD PARTS ILLUSTRATIONS

AN Standard Parts are shown on the following pages and are listed numerically according to their AN number. The equivalent and/or superseding MS numbers are shown where applicable.

AN3 to AN20
BOLT, HEX HEAD

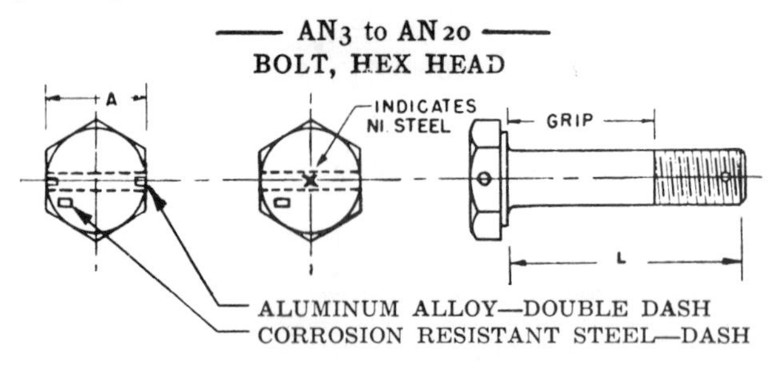

Material: Nickel steel (S.A.E. 2330). Process: heat-treat harden and cadmium plate.

2024-T4 Alum. Process: heat-treat harden and anodic treatment.
Corrosion resistant steel.

Thread: Class 3NF (unless otherwise noted).

DIAMETER—BY PART NUMBER

Part No.	Dia.	A, in.	Part No.	Dia.	A, in.	Part No.	Dia.	A, in.
AN3*	3/16	.375	AN7	7/16	.625	AN12	3/4	1.0625
AN4	1/4	.4375	AN8	1/2	.750	AN14	7/8	1.250
AN5	5/16	.500	AN9	9/16	.875	AN16	1	1.4375
AN6	3/8	.5825	AN10	5/8	.9375			

* SPECIAL NOTE: An AN3 bolt is made to a No. 10-32 screw size (see p. 146) to avoid confusion, since the two measurements are so nearly the same. (No. 10 screw is .190 dia., 3/16 is .187.)

LENGTH—BY DASH NUMBER

Dash No.	L, in.	Dash No.	L, in.	Dash No.	L, in.	Dash No.	L, in.
-3	3/8	-7	7/8	-13	1 3/8	-17	1 7/8
-4	1/2	-10	1	-14	1 1/2	-20	2
-5	5/8	-11	1 1/8	-15	1 5/8	-21	2 1/8
-6	3/4	-12	1 1/4	-16	1 3/4	-22	2 1/4

Additional lengths correspondingly coded in 8ths of an inch.

AN CODE: The material is indicated in the code by a letter preceding the dash number as follows:

Nickel steel .... No letter
Corrosion resistant steel .... Letter C
Aluminum alloy .... Letters DD

The bolt may be supplied with or without the hole in either the shank or the head. This is indicated in code as follows:

UNDRILLED BOLT .... Add A after dash number
DRILLED SHANK ONLY .... No letter required
DRILLED HEAD ONLY .... Add H before dash number and A after dash number
DRILLED HEAD AND DRILLED SHANK .... Add H before dash number

AN21 to AN36

BOLT, CLEVIS

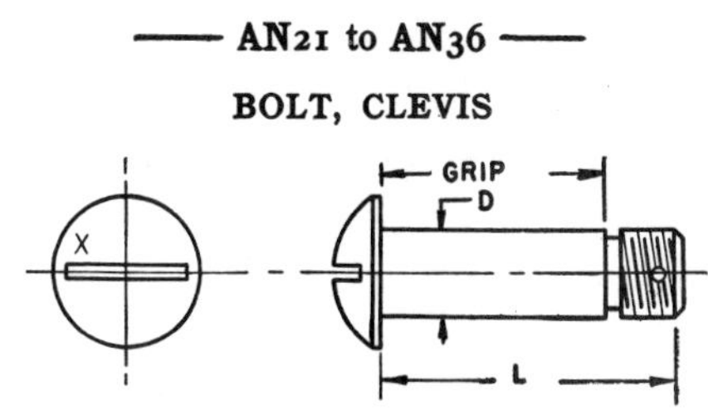

Material: Nickel steel (S.A.E. 2330). Process: heat-treat harden and cadmium plate.

Thread: Class 3NF.

DIAMETER—BY PART NUMBER

Part No.	Size	Part No.	Dia., in.	Part No.	Dia., in.	Part No.	Dia., in.
AN21	6–40	AN24	1/4	AN27	7/16	AN30	5/8
AN22	8–36	AN25	5/16	AN28	1/2	AN32	3/4
AN23	10–32	AN26	3/8	AN29	9/16	AN34	7/8
						AN36	1

LENGTH—BY DASH NUMBER

Dash No.	*L*, in.	Dash No.	*L*, in.	Dash No.	*L*, in.	Dash No.	*L*, in.	Dash No.	*L*, in.
–5	5/16	–9	9/16	–13	13/16	–17	1 1/16	–21	1 5/16
–6	3/8	–10	5/8	–14	7/8	–18	1 1/8	–22	1 3/8
–7	7/16	–11	11/16	–15	15/16	–19	1 3/16	–23	1 7/16
–8	1/2	–12	3/4	–16	1	–20	1 1/4	–24	1 1/2

Additional lengths, on large sizes, to 7 in. are given similarly in 16ths by the dash number.

Example: AN23–12 is clevis bolt size 10–32 and 3/4 in. long.

Description: AN clevis bolts are made from .001 to .002 small to allow a fit in fractional size drill holes. The short thread requires AN320 nut or AN364 self-locking nut. The hole in the threaded end is for a safety cotter; its absence is indicated by the letter A immediately following the dash number in the code.

Example: AN24–15A.

AN42 to AN49

BOLT, EYE

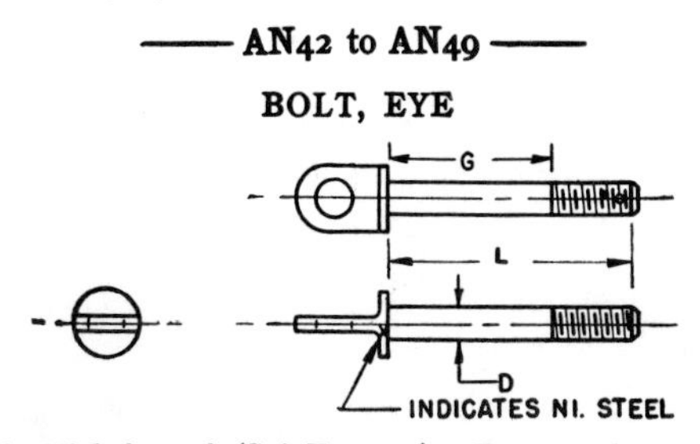

Material: Nickel steel (S.A.E. 2330) Process: heat-treat harden and cadmium plate.

Size: Eye size and shank diameter combination by part number according to chart. Length by dash number in 8ths, same as hex-head bolts.

Part No.	Dia., in.		Clevis strength, lb.	Part No.	Dia., in.		Clevis strength, lb.
	Eye	Shank			Eye	Shank	
AN42	3/16	No. 10–32	1,000	AN47	3/8	7/16–20	8,000
AN43	3/16	1/4–28	2,100	AN48	7/16	1/2–20	11,500
AN45	5/16	5/16–24	4,600	AN49	1/2	9/16–18	15,500
AN46	3/8	3/8–24	6,100				

Shank made 2½ to 4 thousandths undersize and eye is to 10 thousandths oversize.

Example: AN 43—12 is an eyebolt, ¼ in. dia., 3/16 eye, and 1¼ in. long. (Add A for absence of hole.)

—— AN73 to AN81 ——

Superseded by MS20073/74

BOLT, AIRCRAFT-DRILLED HEAD

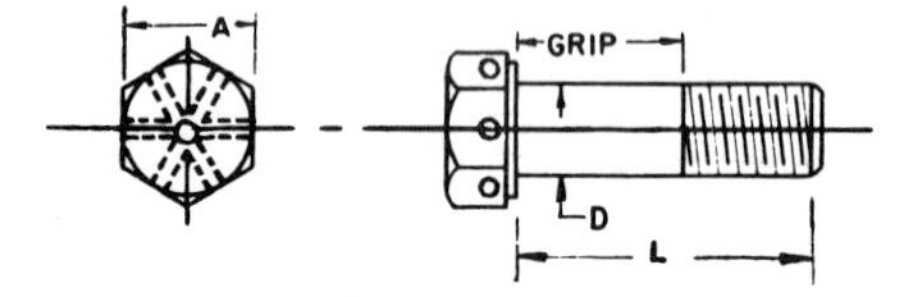

Material: Nickel steel (S.A.E. 2330). Process: heat-treat harden and cadmium plate.

Thread: Class 3NF or class 3NC. (NC indicated by A inserted in code in place of the usual dash.)

Size: A is the same as A for the corresponding sizes AN3 to AN16.

DIAMETER—BY PART NUMBER

Part No.	Dia., in.	Part No.	Dia., in.	Part No.	Dia., in.
AN73*	3/16	AN76	3/8	AN79	9/16
AN74	1/4	AN77	7/16	AN80	5/8
AN75	5/16	AN78	1/2	AN81	3/4

* See special note AN3.

LENGTH—BY DASH NUMBER

Dash No.	*L*, in.	Dash No.	*L*, in.	Dash No.	*L*, in.	Dash No.	*L*, in.
−3	3/8	−7	7/8	−13	1 3/8	−17	1 7/8
−4	1/2	−10	1	−14	1 1/2	−20	2
−5	5/8	−11	1 1/8	−15	1 5/8	−21	2 1/8
−6	3/4	−12	1 1/4	−16	1 3/4	−22	2 1/4

Additional lengths correspondingly coded in 8ths of an inch.

Example: AN74–10 is drilled-head bolt, ¼ dia. and 1 in. long.

Description: The hole in the bolthead is for a safety wire. There is no hole in the threaded end for a cotter.

AN turnbuckle assemblies will be found on older aircraft. However, they have been superseded by MS (Military Standard) turnbuckle assemblies. See page 132 for a description of MS turnbuckle assemblies.

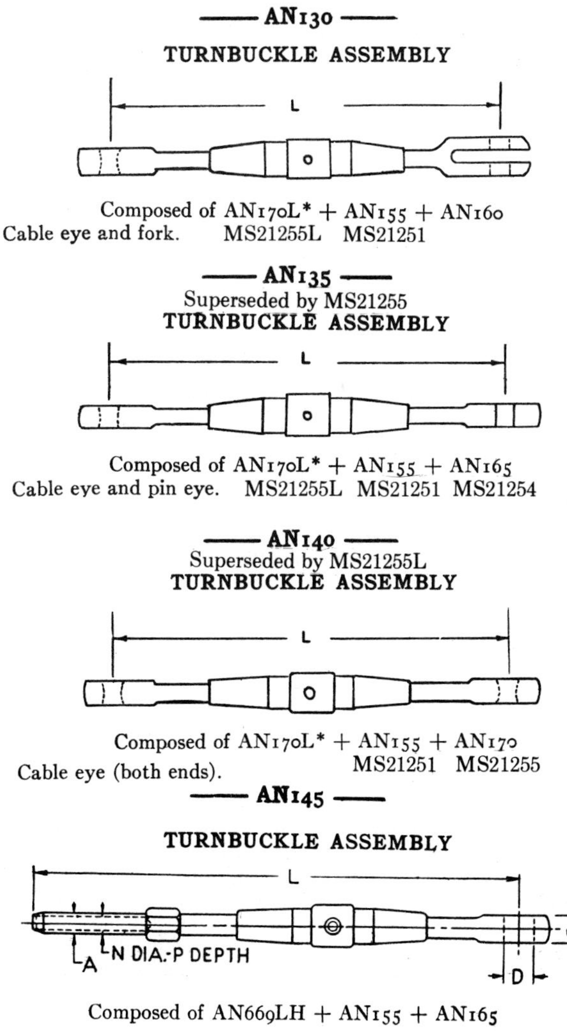

— AN130 —

TURNBUCKLE ASSEMBLY

Composed of AN170L* + AN155 + AN160
Cable eye and fork. MS21255L MS21251

— AN135 —

Superseded by MS21255

TURNBUCKLE ASSEMBLY

Composed of AN170L* + AN155 + AN165
Cable eye and pin eye. MS21255L MS21251 MS21254

— AN140 —

Superseded by MS21255L

TURNBUCKLE ASSEMBLY

Composed of AN170L* + AN155 + AN170
Cable eye (both ends). MS21251 MS21255

— AN145 —

TURNBUCKLE ASSEMBLY

Composed of AN669LH + AN155 + AN165

Swaging terminal and pin eye.

* Left-hand thread.

AN146

TURNBUCKLE ASSEMBLY

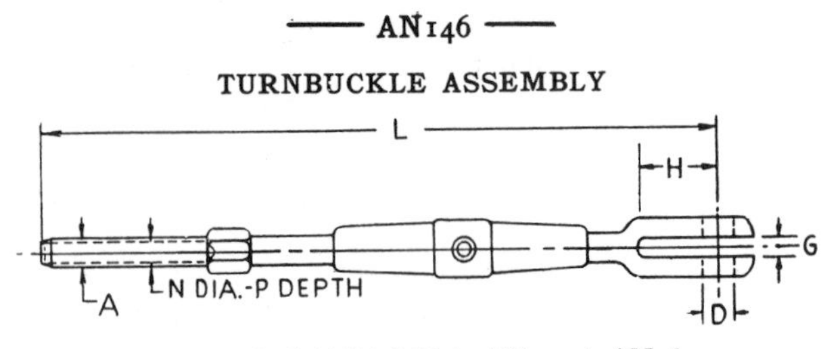

Composed of AN669LH + AN155 + AN161
Superseded by MS21252

Swaging terminal and fork.

AN147

TURNBUCKLE ASSEMBLY

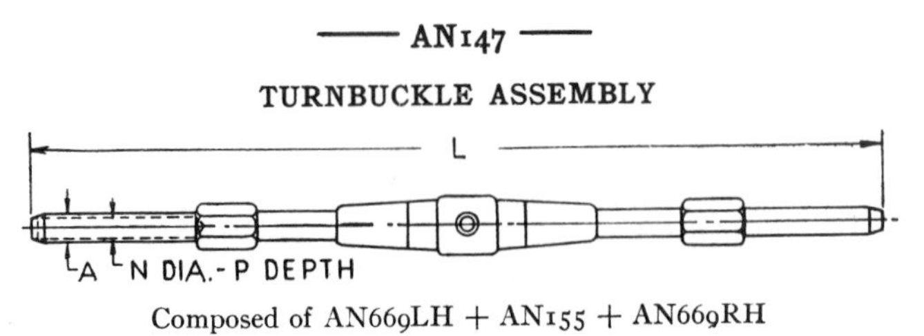

Composed of AN669LH + AN155 + AN669RH

Swaging terminal, both ends.

AN150

Superseded by MS21252L

TURNBUCKLE ASSEMBLY

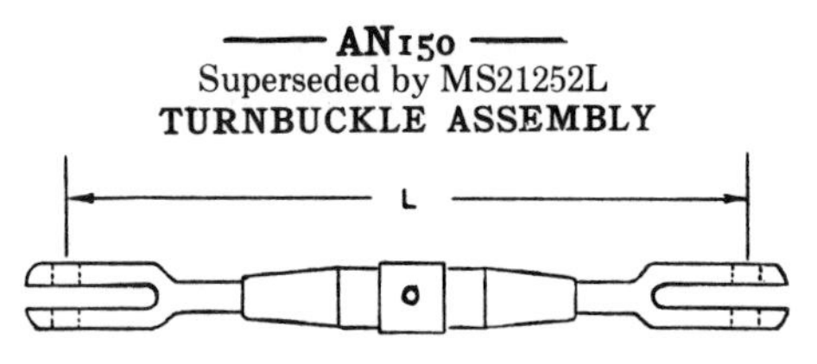

Composed of AN161L* + AN155 + AN161

Fork (both ends). Use AN111 bushing.

Description: The turnbuckle length equals the length of the barrel plus the effective lengths of both ends with the ends screwed clear in. Notice that the dash numbers of the ends are the same as the dash numbers for the barrels or turnbuckle assemblies in which they fit. Notice also that the lockwire tends to tighten the turnbuckle (Fig. 14, p. 64). These turnbuckles are, as already noted, combinations of the basic components described below. All turnbuckles and components are in two general classifications: long and short indicated by L or S in code.

* Left-hand thread.

The various sizes are indicated by dash numbers designating the rated pounds strength. The short types come in 8-, 16-, 21-, 32-, and 46-hundred pound strengths, while the long types come in 16-, 21-, 32-, 46-, 61-, 80-, 125-, and 175-hundred pound strengths. Class 3NF threads as follows are used on all turnbuckle parts:

Dash No.	Thread	Dash No.	Thread	Dash No.	Thread	Dash No.	Thread
-8	6-40	-21	12-28	-46	5/16-24	-125	7/16-20
-16	10-32	-32	1/4-28	-61 + -80	3/8-24	-175	1/2-20

Examples: AN130-8S is turnbuckle assembly, cable eye and fork, short, 8 hundred pounds strength. AN135-21L is turnbuckle assembly, cable eye and pin eye, long, 21 hundred pounds strength. AN140-32S is turnbuckle assembly, cable eyes, short, 32 hundred pounds strength. AN150-16S is turnbuckle assembly, fork ends, short, 16 hundred pounds strength.

AN155

TURNBUCKLE BARREL

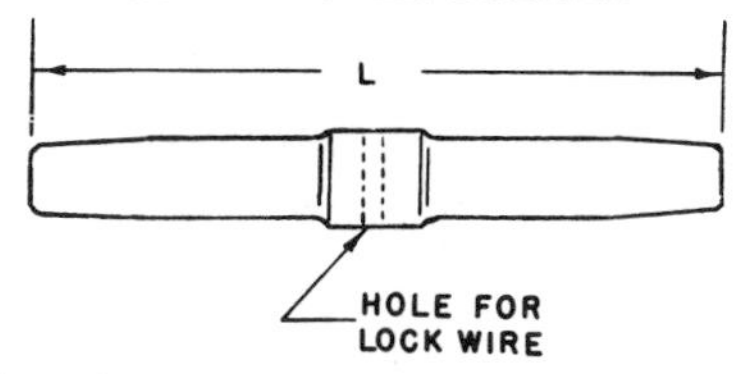

Material: Brass.

Size: Lengths as follows: -8S to -46S incl., 2¼ in.; -16L to -80L incl., 4 in.; -125L and -175L, 4¼ in.

AN160

TURNBUCKLE FORK

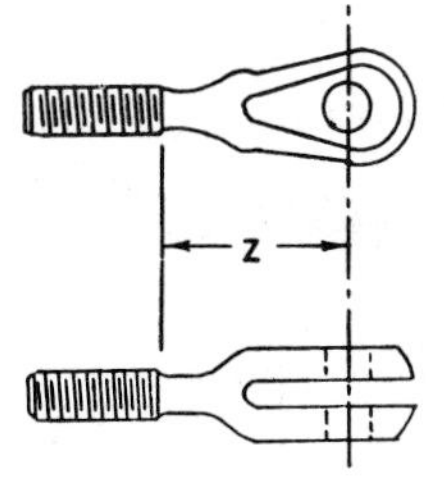

Material: Steel. Process: heat-treat and cadmium plate.

Description: Right-hand thread. *Z*, or effective length, as follows: -8S to -46S incl., 1⅛ in.; -16L to -80L incl., 2 in.; -125L, 2⅜; -175L, 2⅝ in.

AN161
MS21252
TURNBUCKLE FORK

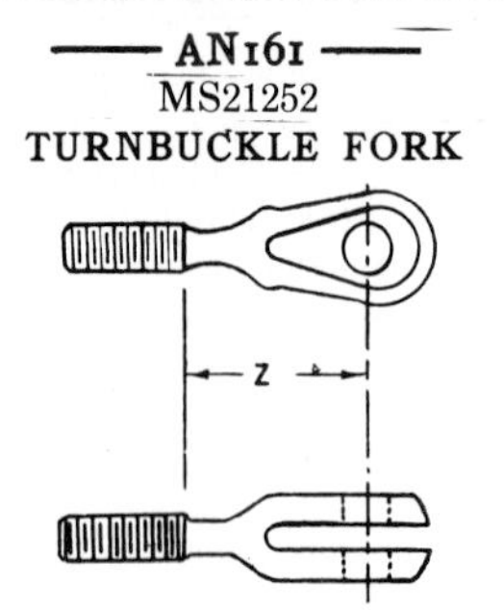

Material: Steel. Process: heat-treat and cadmium plate.

Description: Right- or left-hand thread, indicated by R or L. *Z*, or effective lengths, as follows:

Dash No.	*Z*, in.	Dash No.	*Z*, in.	Dash No.	*Z*, in.	Dash No.	*Z*, in.
–16S	1 1/8	–46S	1 9/32	–32L	2 7/64	–80L	2 5/16
–21S	1 5/32	–16L	2	–46L	2 5/32	–125L	2 7/16
–32S	1 7/32	–21L	2 1/32	–61L	2 9/32	–175L	2 11/16

AN165
MS21254
TURNBUCKLE EYE FOR PIN

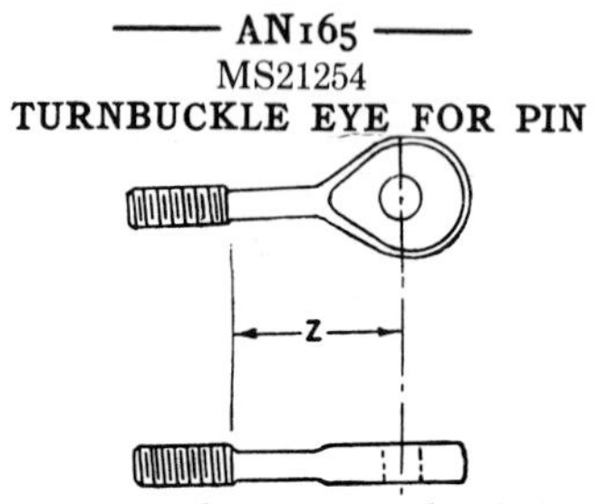

Material: Steel. Process: heat-treat and cadmium plate.

Description: Right-hand thread. *Z*, or effective lengths, same as AN160.

AN170
MS21255
TURNBUCKLE EYE FOR CABLE

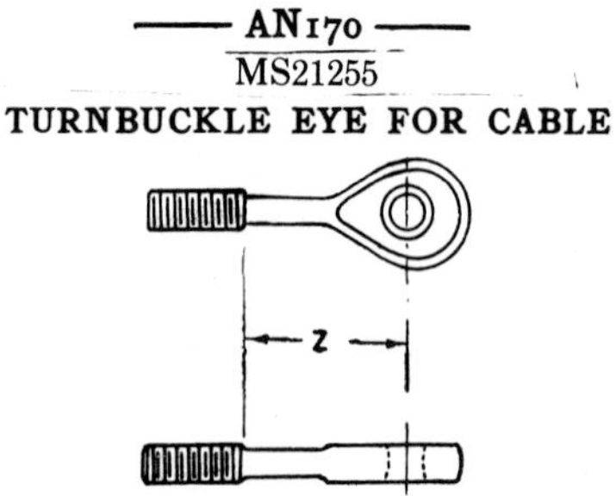

Material: Steel. Process: heat-treat and cadmium plate.

Description: Right- or left-hand thread, indicated by R or L. *Z*, or effective length, same as AN160.

—— AN 173 to AN 186 ——

BOLT — CLOSE TOLERANCE

This marking on head of close tolerance bolts.

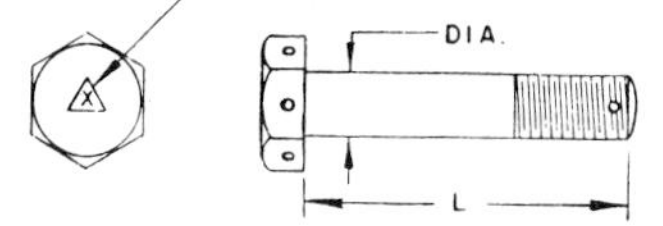

Description: These bolts are made of nickel steel (S.A.E. 2330). Head and threads are cadmium-plated. Shank is ground with a tolerance of .0005 of an inch. Diameter is ±.0005. The shank is greased after grinding. Thread is NF.

Size: Diameter given by part number. Length is given in 8ths by the second dash number according to Table 4, page 144 . These bolts may be obtained with or without drilled shank and with or without drilled head. (Code like AN3 to AN20)

—— AN310 ——

NUT, AIRCRAFT—CASTLE

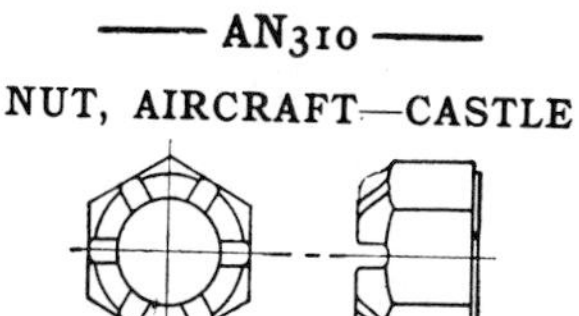

Material: Steel. Process: heat-treat harden and cadmium plate.
2024-T4 Alum. indicated by D preceding dash number.
Corrosion resistant steel indicated by C preceding dash number.

Thread: Class 3 NF.

Size: Made in sizes 3/16 to 1 in., indicated in 16ths of an inch by the dash number. Fits AN hex-head bolts of same designated size.

Example: AN310—5 is AN castle nut made of steel and fits a 5/16 AN bolt.

—— AN315 ——

NUT, AIRCRAFT—PLAIN

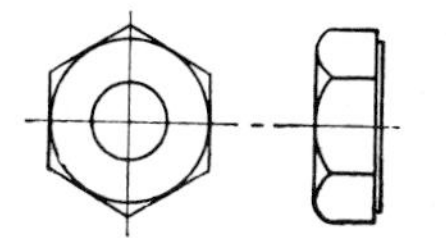

Material, Thread, and Sizes: Same as AN310, except that thread is also left-hand indicated by L or R following dash number and is made in size 640

Examples: AN315D5L is plain aircraft nut made of dural. Fits a 5/16 bolt and has a left-hand thread. AN315D5R is same nut with a right-hand thread.

AN316

NUT, AIRCRAFT—CHECK

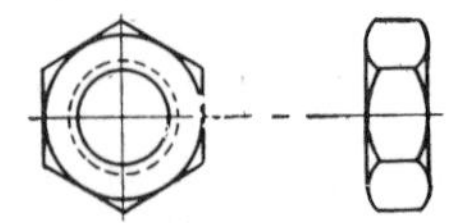

Material: Steel. Process: heat-treat harden and cadmium plate.
Thread: Class 3 NF, left or right indicated by L or R.
Sizes: Made in sizes 1/4 to 1 in. incl. indicated in 16ths of an inch by this dash number.
Description: Much thinner than AN315.
Examples: AN316–6R is aircraft check nut to fit a 3/8 bolt with right-hand thread. AN316–6L is same nut with left-hand thread.

AN320

NUT, AIRCRAFT—SHEAR

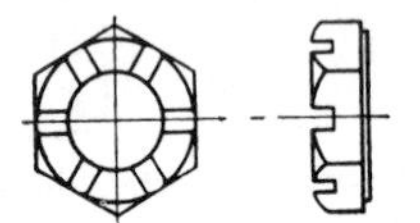

Material and Thread: Same as AN310.
Size: Same as AN310 except –1 is size 6–40, –2 is size 8–36, and –20 is 1 1/4–12.
Description: Much thinner than AN310 as illustration indicates.
Examples: AN320–4 is aircraft shear nut made of steel and fits a 1/4-in. AN bolt. AN320D4 is same nut made of aluminum alloy.

AN340

NUT, MACHINE SCREW—HEX (COARSE THREAD)

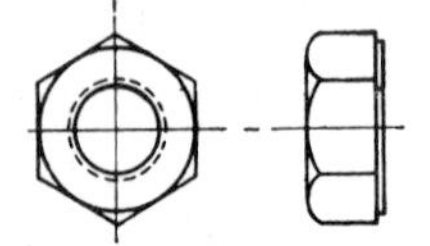

Material: Steel, cadmium-plated.
Corrosion resistant steel
Brass, commercial.
2024-T4 Alum., anodized.

Thread: Class 2 NC.

Size: Machine screw sizes; steel, 2 to 416; brass, 2 to 10; dural, 6 to 616. Refer to dash number table, page 144

Examples: AN340DD6 is machine screw hex nut made of 2024 aluminum to fit a size 6-32 machine screw. AN340B6 is same nut made of brass. AN340-6 is same nut made of steel. AN340C6 is same nut made of corrosion resistant steel.

—— AN345 ——

NUT, PLAIN HEX (FINE THREAD)

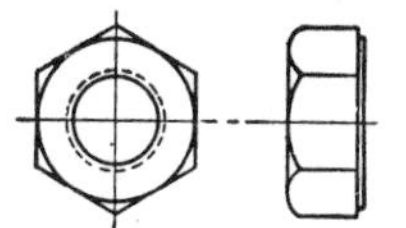

Material: Carbon steel, commercial, cadmium-plated.
Brass, commercial, indicated by B in code.
Corrosion-resistant steel, passivated. Indicated by C in code.
2024-T4 aluminum alloy, anodized. Indicated by DD in code.

Thread: Class 2 NF.

Sizes: Machine screw, dural No. 10 and 416; brass 0 to 10 incl.; carbon steel 0 to 416 incl.; and corrosion-resistant 0 to 416 incl.

Examples: AN345C4 is machine screw hex nut (fine thread) made of stainless (corrosion-resistant) steel, size 4. AN345B4 is the same screw made of brass. AN345DD4 is the same screw made of 2024-T4 Alum AN345–4 is the same screw made of carbon steel.

—— AN350 ——

NUT, WING

Material: Steel, cadmium-plated, or brass.

Thread: Class 2 NF.

Sizes: 6, 8, and 10 machine screw sizes coded according to dash number Table 1, page 144 and ¼, 5/16, 3/8, 7/16, and ½ indicated by dash number giving diameter of bolt that it fits in 16ths of an inch.

Examples: AN350–1032 is wing nut made of steel and fits a 10–32 machine screw. AN350B4 is wing nut made of brass and fits a ¼-in. bolt.

—— AN355 ——

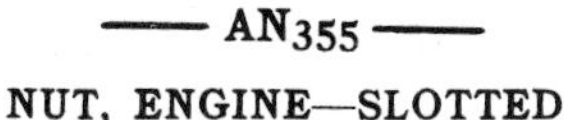

NUT, ENGINE—SLOTTED

These nuts are designed specifically for engines and are *not* to be used on aircraft.

Material: Steel. Process: heat-treat harden. Spec. AN-QQ-S-689, AN-QQ-690, AN-QQ-687. Spec. 29-26 except material.

Thread: Class 3 NF.

Size: Made in 3/16 to 3/4 sizes indicated by dash number in 16ths giving diameter of bolt that it fits (-3 is 10-32).

Example: AN355-5 is slotted engine nut made of steel and fits a 5/16 bolt.

— AN356 —

NUT, LOCK (PALNUT)

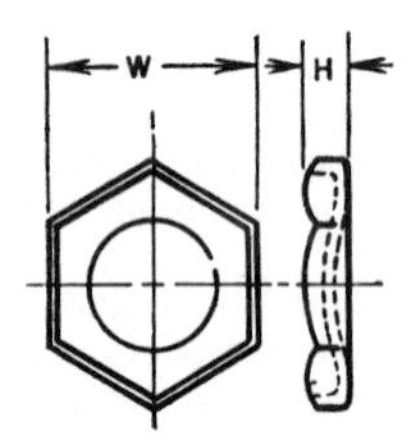

Material: Steel. Process: cadmium plate (Spec. AN-QQ-P-421).

Description: *Not to be used in aircraft structures.* Is made in National Coarse (No. 4-36 to 3/4-10) and National Fine (No. 10-32 to 3/4-16) threads.

Sizes: Sizes are indicated according to thread Table 1, page 140.

Example: AN356-524 is AN palnut 5/6 in. diameter with 24 threads per inch.

— AN360 —

NUT, ENGINE—PLAIN

These nuts are designed specifically for engines and are *not* to be used on aircraft.

Material, Thread, and Size: Same as AN355 except finish is black rustproof (Spec. 57-0-2).

Example: AN360-5 is plain engine nut made of steel and fits a 5/16 bolt.

— AN 362 —

NUT, SELF-LOCKING PLATE (HIGH TEMPERATURE)

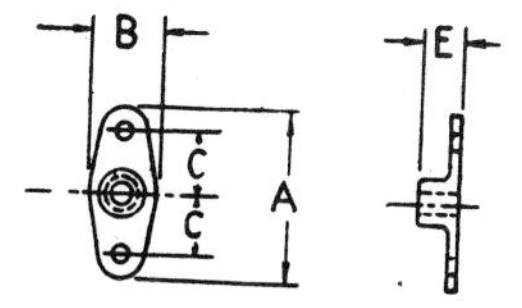

Material: Steel.
Corrosion resistant steel
Thread: NF and NC.
Sizes: No. 6 machine screw through ⅜ dia. Coded according to dash number, Table 1 on page 144.

Code: F — Steel
C — Corrosion resistant steel
WC — Weldable corrosion resistant steel

— AN363 —

NUT, SELF-LOCKING (HIGH TEMPERATURE)

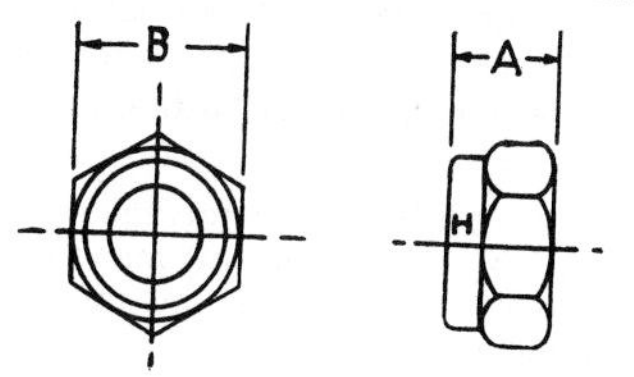

Material: Steel .Corrosion resistant steel or brass.
Thread: NF and NC.
Sizes: No. 10 machine screw through ¾ dia. Coded according to Table 1 on page 144. Shall not be used where temperature exceeds 650°F.
Description: Similar in appearance to AN365.

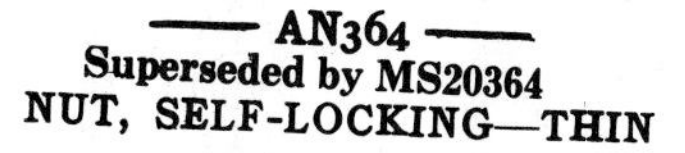

— AN364 —

Superseded by MS20364

NUT, SELF-LOCKING—THIN

Material: Steel. Process: cadmium plate.
Brass.
Dural 24S, heat-treat hardened and anodized.

Thread: Class 2 NF on 6–40 and 836. Class 3 NF on 10–32 through 1 in.

Size: Steel 6–40 through 1 in., brass 8–36 through 1 in. and dural 10–32 through 1 in. All sizes designated according to dash number Table 1 on page 144.

Examples: AN364-1032 is thin self-locking nut made of steel to fit a 10-32 machine screw. AN364B1032 is same nut made of brass. AN364D1032 is same nut made of dural.

Description: Add A after dash number for non-metalic inserts. Add C after dash number for all metal.

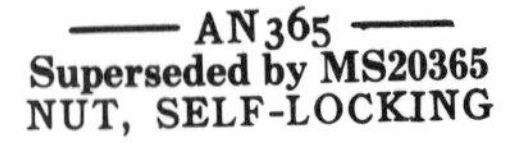

—— AN365 ——
Superseded by MS20365
NUT, SELF-LOCKING

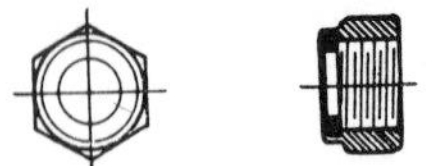

Material: Same as AN364.

Thread: Class 2 NF or NC on machine screw sizes 4, 6, and 8. Class 3 NF or NC on machine screw size 10 through ½ in. dia.

Size: Steel machine screw size 4 through ½ dia.; brass fine thread size 8 through ½ in. dia. and in coarse thread size 4 through ½ in. dia.; dural fine thread size 10 through ½ in. dia., and coarse thread size 4 through ⅜ dia. All sizes designated according to dash number table 1, page 144.

Examples: AN365-1032 is self-locking nut made of steel and fits a 10-32 screw. AN365B524 is self-locking nut made of brass and fits a 5/16 fine thread bolt. AN365D832 is self-locking nut made of dural and fits an 8-32 machine screw.

Description: Add A after dash number for non-metalic inserts. Add C after dash number for all metal.

—— AN366 ——
Superseded by MS21048
NUT, PLATE

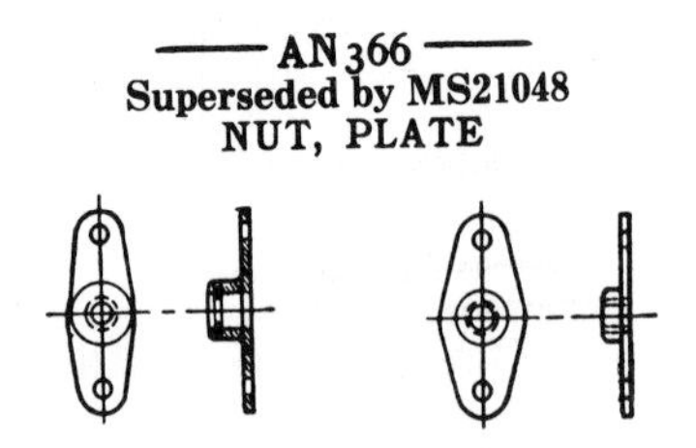

Material: Steel. Process: cadmium plate.
2024-T4 Alum: Process: heat-treat harden and anodize.

Thread: Class 2 NF.

Size: No. 8, No. 10, and ¼ in. except self-locking steel, which also has No. 6. All sizes designated according to Table 1, page 144.

Description: Self-locking nuts shall not be used where the temperature exceeds 250°F. To be riveted in place where nut is inaccessible to a wrench. Self-locking type indicated in code by F preceding the dash number. Add A after dash number for non-metalic inserts. Add C after dash number for all metal.

Examples: AN366-836 is plain plate nut made of steel, size 8-36. AN366DF420 is self-locking plate nut made of dural and fits a ¼ in. dia. coarse thread bolt. AN366F428 is same plate nut made of steel to fit a fine thread bolt.

—— AN380 ——
Superseded by MS24665
PIN, COTTER

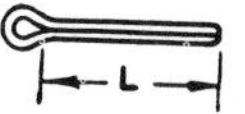

Material: Low carbon steel. Process: cadmium plate.
Size: See following table:

First dash No.	Dia., in.
−2	1/16
−3	3/32
−4	1/8
−5	5/32
−6	3/16
−8	1/4

Second dash No.	Length, in.	Second dash No.	Length, in.
−2	1/2	−8	2
−3	3/4	−10	2½
−4	1	−12	3
−5	1¼	−14	3½
−6	1½	−16	4
−7	1¾		

—— AN381 ——
Superseded by MS24665
PIN, COTTER

Description: Same as above except made of corrosion resistant steel.

—— AN385 ——
PIN, TAPER

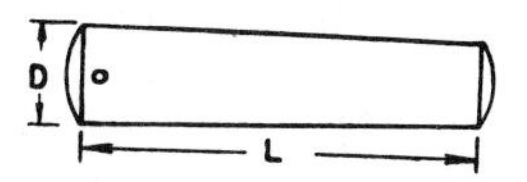

Material: Steel, plain carbon and alloy steel.

Sizes

First dash No.	Dia. large end, in.	First dash No.	Dia. large end, in.	First dash No.	Dia. large end, in.
−60	.078	−20	.141	−3	.219
−50	.094	−10	.156	−4	.250
−40	.109	−1	.172	−5	.289
−30	.125	−2	.193	−6	.341

Second dash No.	Length, in.	Second dash No.	Length, in.	Second dash No.	Length, in.
−3	3/8	−8	1	−18	2 1/4
−4	1/2	−10	1 1/4	−20	2 1/2
−5	5/8	−12	1 1/2	−22	2 3/4
−6	3/4	−14	1 3/4	−24	3
−7	7/8	−16	2		

Add H before dash Number for drilled head.
Add A before dash number for alloy steel.
Add P before second dash number for cadmium plate.

Examples: AN385–30–6 is taper pin, size 3/0, Morse taper (1/4 in. per foot), 3/4 in. long. AN385–3–12 is taper pin, size 3, Morse taper, 1 1/2 in. long. Use Morse standard taper pin reamer, same designated size as pin. Use drill as specified or, when not specified, .003 to .005 smaller than small end of pin.

—— AN386 ——

PIN, TAPER—THREADED

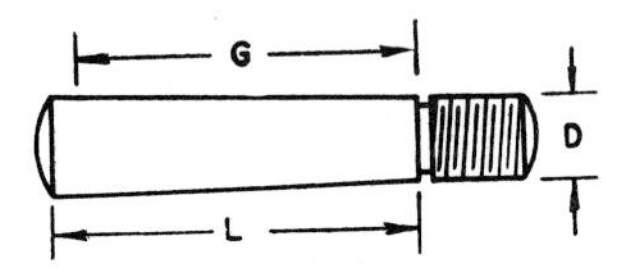

Material: Steel. Process: heat-treat harden and cadmium plate.
Thread: Class 3 NF.

Sizes

First dash No.	Dia., in.	Thread	First dash No.	Dia., in.	Thread	First dash No.	Dia., in.	Thread
−1	.2052	10–32	−5	.4552	3/8–24	−9	.9052	3/4–16
−2	.2052	10–32	−6	.5052	7/16–20	−10	1.0502	7/8–14
−3	.3172	1/4–28	−7	.6052	1/2–20	−10A	1.1527	7/8–14
−4	.3552	5/16–24	−8	.7552	9/16–18	−11	1.2552	7/8–14
L = grip + 1/8			*L* = grip + 3/16			*L* = grip + 1/4		

Second dash No.*	Grip, length,† in.	Second dash No.*	Grip, length,† in.	Second dash No.*	Grip, length,† in.
−6	3/4	−13	1 5/8	−19	2 3/8
−7	7/8	−14	1 3/4	−20	2 1/4
−8	1	−15	1 7/8	−21	2 5/8
−9	1 1/8	−16	2	−22	2 3/4
−10	1 1/4	−17	2 1/8	−23	2 7/8
−11	1 3/8	−18	2 1/4	−24	3
−12	1 1/2			−25	3 3/8

* A following second dash number indicates no cotter pin hole.
† Additional lengths to 6 in. are correspondingly indicated in 8ths.

Examples: AN386–6–12 is threaded taper pin, Brown & Sharpe taper (1/2 in. per foot except No. 10 and No. 10A, on which taper is .5161 in. per foot), size 6 with a 1 1/2 in. grip length. AN386–612A is same pin without cotter pin hole. Use Brown & Sharpe taper reamer with same number size as pin. Use drill as specified or, when not specified, 3 to 5 thousandths smaller than small end of taper.

AN392 to AN406
Superseded by MS20392
PIN, FLAT HEAD

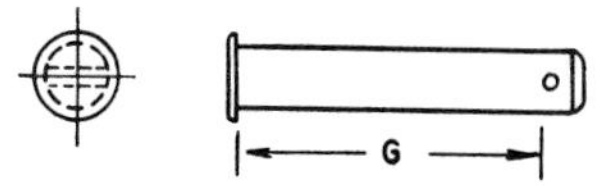

Material: Steel. Process: heat-treat harden and cadmium plate.

SIZES

Part No.	Dia.,* in.	Part No.	Dia.,* in.	Part No.	Dia.,* in.
392	1/8	396	3/8	400	5/8
393	3/16	397	7/16	402	3/4
394	1/4	398	1/2	404	7/8
395	5/16	399	9/16	406	1

Dash No.	Grip length†	Dash No.	Grip length†	Dash No.	Grip length†
–7	7/32	–17	17/32	–29	29/32
–9	9/32	–19	19/32	–31	31/32
–11	11/32	–21	21/32	–33	1 1/32
–13	13/32	–23	23/32	–35	1 3/32
–15	15/32	–25	25/32	–37	1 5/32
		–27	27/32		

* 0 to .002 under indicated size.
† Additional grip lengths to 4 9/32 given similarly in odd 32nds.

Example: AN394–25 is flathead pin 1/4 dia. and 25/32 grip length.

AN415

PIN, LOCK

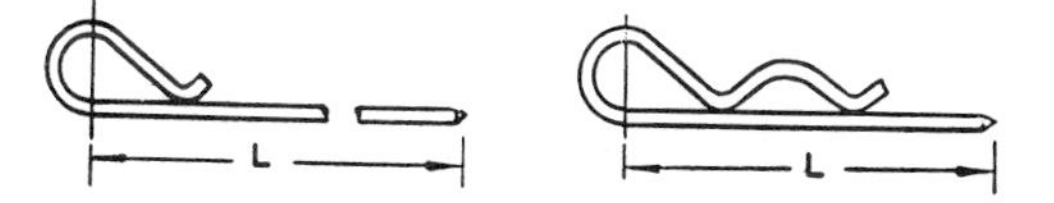

Material: Corrosion resistant steel
Size: Length of AN415 given in increments of 1 in. by dash number.
Example: AN415–3 is lock pin 3 in. long.

AN416

PIN, RETAINING—SAFETY

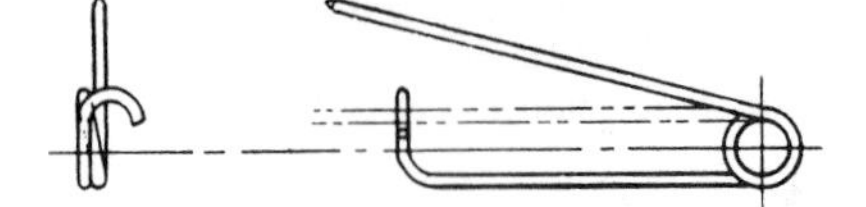

Material: Steel wire. Process: cadmium-plate.

SIZES

Dash No.	A, in.	B, in.	D, in.
−1	1 1/16	1 3/8	.051
−2	3/4	1 5/16	.041

AN426
MS20426 (100°)
RIVET, 100° COUNTERSUNK HEAD—ALUMINUM ALLOY

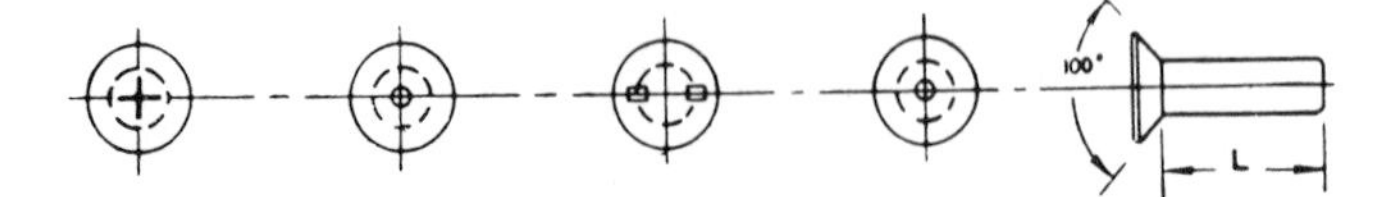

Material: Alloy 5056-H12 (B), ¼ hard.
Alloy 2117-T4 (AD), heat-treat hardened.
Alloy 2024-T4 (DD), heat-treat hardened.
Alloy 2017-T4 (D), heat-treat hardened.

Size: Indicated by dash numbers according to Table 4, page 144.

Example: AN426B3-5 is 100-deg. countersunk-head rivet, made of 5056-H12, 3/32 dia. and 5/16 long. AN426AD3-5 is same rivet made of 2117-T4, AN426DD3-5 is same rivet made of 2024-T4. AN426D3-5 is same rivet made of 2017-T4

AN427
MS20427
RIVET, 100° COUNTERSUNK

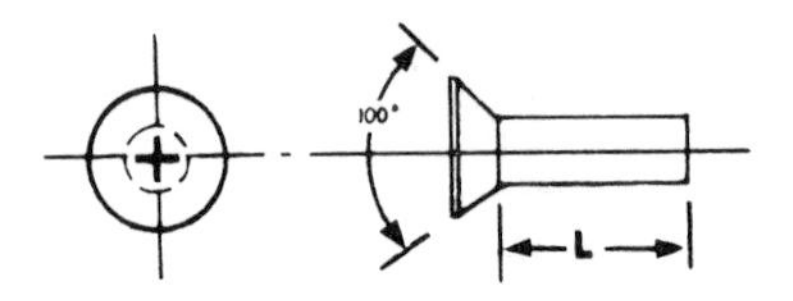

Material: Steel, monel, corrosion resistant steel and copper.

Size: Indicated by dash numbers according to table 4, page 144.
Add before the dash number:—
F—for corrosion resistant steel
M—for monel
C—for copper
Add after the dash number:—
C—for cadmium plated carbon steel
Z—for zinc plated
U—for unplated
The monel rivet is indicated by the triangle on the rivet head.

—— AN430 ——
MS20430
RIVET, ROUND HEAD—ALUMINUM ALLOY *

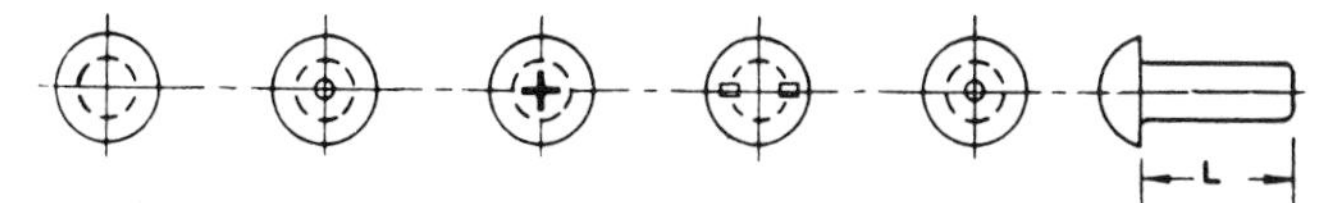

*These rivets have been superseded by AN470 for most applications.
Material: Alloy 1100-H14 (A), not applicable for new design.
Alloy 2117-T4 (AD), heat-treat hardened.
Alloy 5056-H12 (B), ¼ hard.
Alloy 2024-T4 (DD), heat-treat hardened.
Alloy 2017-T4 (D), heat- reat hardened.
Size: Indicated by dash numbers according to Table 4, page 144.
Example: AN430A3-5 is roundhead rive made of 2S½H, 3/32 dia., 5/16 long. AN430AD3-5 is same rivet made of 2117-T4. AN430B3-5 is same rivet made of 5056-H12. AN430DD3-5 is same rivet made of 2024-T4. AN430D3-5 is same rivet made of 2017-T4.

—— AN442 ——

RIVET, FLAT HEAD—ALUMINUM *

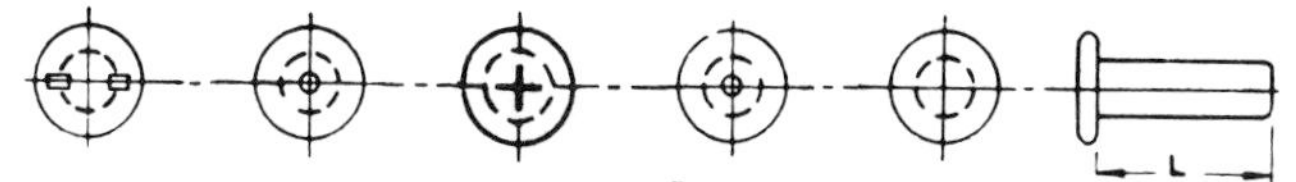

*These rivets have been superseded by AN470 for most applications.
Material: Alloy 2017-T4 (D), heat-treat hardened;
Alloy 2024-T4 (DD), heat-treat hardened;
Alloy 5056-F
Alloy 1100-H14 (A), half hard.
Alloy 2117-T4 (AD), heat-treat hardened.
Size: Indicated by dash numbers according to Table 4, page 144.
Example: AN442A3-5 is flathead rivet made of 1100-H14, 3-32 dia., 5/16 long. AN442AD3-5 is same rivet made of 2117-T4

—— AN456 ——

RIVET, BRAZIER HEAD—ALUMINUM ALLOY *

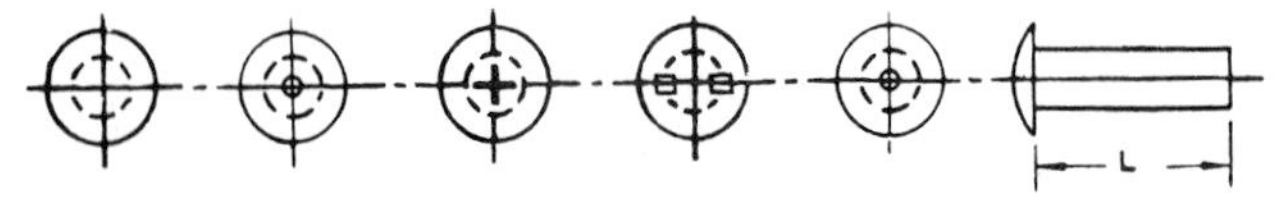

*These rivets have been superseded by AN470 for most applications.

Material: Alloy 2017-T4 (D), 2117-T4 (AD), 2024-T4 (DD), 5056-F (B).
Size: Inidcated by dash numbers according to Table 4, page 144.
Example: AN456AD3-5 is brazier head rivet, made of 2117-T4, 3/32 dia., 5/16 long. AN456D3-5 is same rivet made of 2017-T4.

—— AN470 ——
MS20470
RIVET — UNIVERSAL HEAD

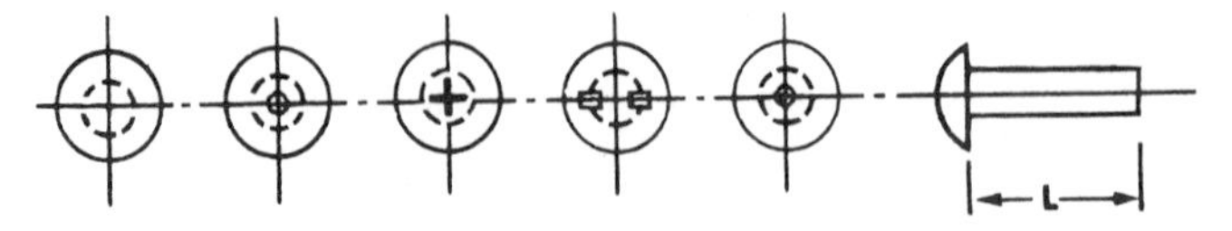

Material and **Size** of this rivet are the same as AN430. This rivet is a replacement rivet to suit the requirements of AN430, AN442, and AN456.

—— AN481 ——

ROD END, CLEVIS

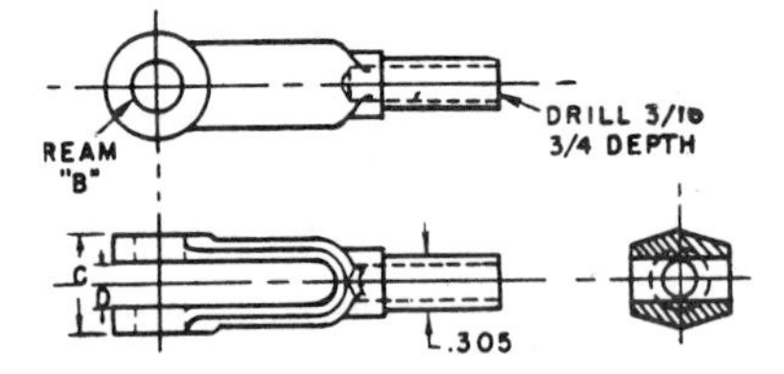

Material: Steel forging (Spec. 57-107-20). Finish: heat-treat normalize. Cadmium plate indicated by P following dash number. Unplated parts to be finished at assembly.

SIZE

Dash No.	*B*, in.	*C*, in.	*D*, in.	Dash No.	*B*, in.	*C*, in.	*D*, in.
−1	.250	1/2	17/64	−3	.1875	1/2	17/16
−2	.250	5/16	9/64	−4	.1875	5/16	9/64

AN486 same as AN481 except that the end opposite the clevis has Y4-28 NF3 internal thread.

—— AN665 ——

TERMINAL, TIE ROD

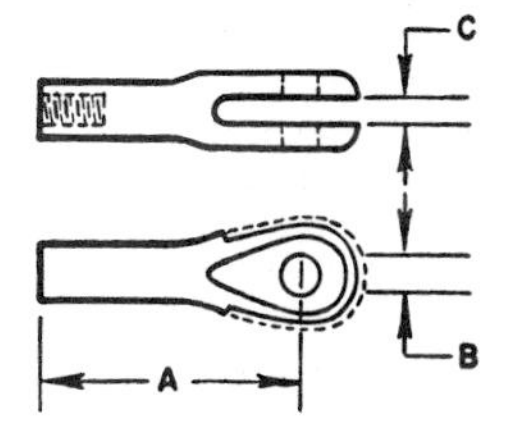

Types: Threader—clevis.
Material: Nickel steel. Process: heat-treat.
Threads: NF–3. Left and right hand, denoted by L or R following dash number.

SIZES

Dash No.	Tie rod strength, lb.	Tap T	A, in.	B, in.	C, in.
–10	1,200	6–40	1 5/16	.188	.109
–21	2,400	10–32	1 17/32	.188	.150
–34	4,200	1/4–28	1 13/16	.250	.203
–46	4,600	5/16–24	1 7/8	.313	.203
–61	6,900	5/16–24	2	.375	.203
–80	10,000	3/8–24	2 1/4	.375	.266
–115	13,700	7/16–20	2 1/2	.438	.344
–155	18,500	1/2–20	2 13/16	.500	.406
–202	24,000	9/16–18	3 1/8	.563	.453
–247	29,500	5/8–16	3 3/8	.625	.516
–430	42,000	3/4–16	4 1/8	.750	.656
–580	58,000	7/8–14	4 7/8	.875	.781
–760	76,000	1–14	5 3/4	1.	.906

Examples: AN665–61L. AN665–80R.

——AN500——

SCREW, FILLISTER HEAD—DRILLED OR PLAIN HEAD (COARSE THREAD)

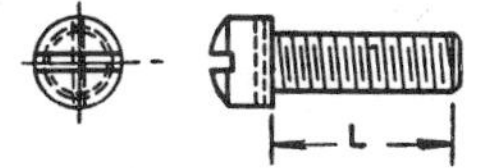

Material: Steel (S.A.E. 1120). Process: cadmium plate.
Stainless steel (corrosion-resistant 18–8) passivated; C in code.
Brass (commercial); B in code.

Thread: Class 2 NC.

Size: Machine screw size or diameter of longer sizes is given by first dash number according to screw size Table 2, page 144. Sizes are No. 2 through 3/8 dia. Length in 16ths of an inch is given by second dash number according to Table 3, page 144.

Examples: AN500-10-14 is fillister-head screw, made of carbon steel, size No. 10-24, and 7/8 in. long. AN500C10-14 is the same screw made of stainless steel. AN500B10-14 is the same screw made of brass.

Description: Drilled head is indicated by A immediately following the part number.

—— AN501 ——

SCREW, FILLISTER HEAD

Drilled or plain head (fine thread). Identical with AN500 except that thread is class 2 NF. Sizes are No. 0 through 3/8 dia.

—— AN502 ——

SCREW, AIRCRAFT—FILLISTER HEAD, DRILLED (FINE THREAD)

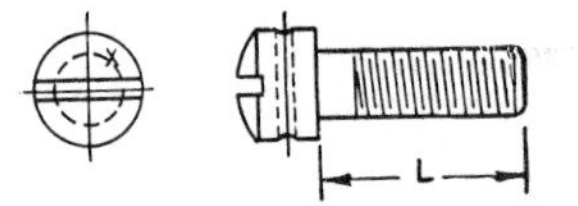

Material: Steel (S.A.E. 2330) (S.A.E. 3140 or 6150 optional). Process: heat-treat harden and cadmium plate.

Thread: Class 3 NF.

Size: Machine-screw size or diameter of large sizes is given by first dash number according to Table 2, page 144. Sizes are 6, 8, 10, 1/4, and 5/16. Length in 16ths of an inch is given by second dash number according to Table 3, page 144.

Example: AN502–8–10 is fillister-head screw, size 8–36 and 5/8 in. long.

—— AN503 ——

SCREW, AIRCRAFT—FILLISTER HEAD, DRILLED (COARSE THREAD)

Identical with AN502 except that thread is class 3 NC.

—— AN505 ——

SCREW, FLAT HEAD (COARSE THREAD)

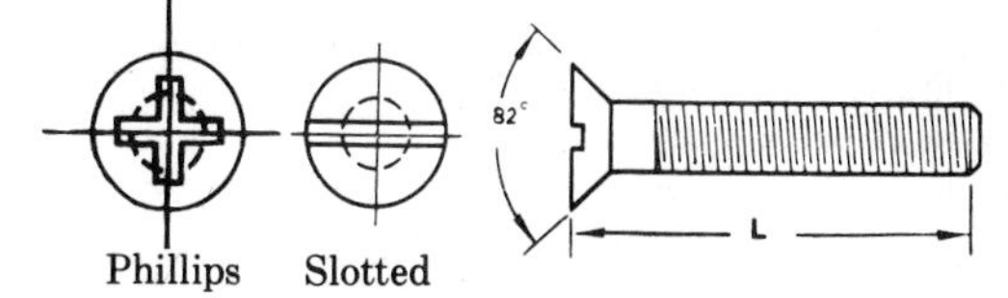

Material: Steel (S.A.E. 2330). Process: cadmium plate.
Brass.
Aluminum alloy Process: heat-treat harden, anodize. D in code.
Thread: Class 2 NC.
Size: Machine screw size or diameter is given by first dash number according to Table 2, page 144. Sizes are 2 through 3/8 dia. Length in 16ths of an inch are given by second dash number according to Table 3, page 144.
Examples: AN505–10–16 is flathead (82 deg countersunk) machine screw, size 10–24, 1 in. long, made of steel. AN505B10–16 is same screw made of brass. AN505D10–16 is same screw made of aluminum alloy. Add C for corrosion resistant steel. Add R between dash numbers for recessed head.

—— AN509 ——

100° FLUSH HEAD SCREW

(Same as above except for head angle.)

—— AN510 ——

SCREW, FLAT HEAD (FINE THREAD)

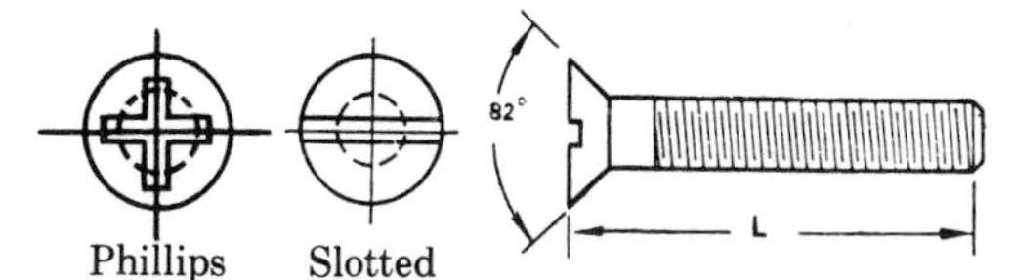

Material: Steel (S.A.E. 1120). Process: cadmium-plate.
Corrosion-resistant steel, passivated. C in code.
Brass. B in code.
Aluminum alloy, "Inactive for design." D in code.
Thread: Class 2 NF.
Sizes: No. 0 through 1/4 in. dia. coded same as AN505
Example: AN510C10–16 is corrosion-resistant steel.
Add R between dash numbers for recessed head.

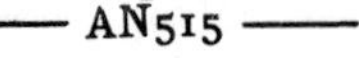

—— AN515 ——

SCREW, ROUND HEAD (COARSE THREAD)

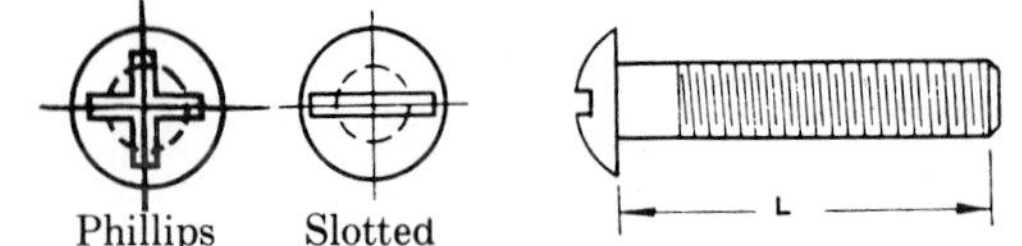

Phillips Slotted

Material: Steel (S.A.E. 1120). Process: cadmium plate.
Brass.
Aluminum alloy. Process: heat-treat harden and anodize.

Thread: Class 2 NC.

Size: Machine screw size or diameter of larger sizes are given by first dash number according to Table 2, page 144. Sizes are No. 2 through 3/8 dia. Length in 16ths of an inch in given by second dash number according to Table 3, page 144.

Example: AN515D10–16.

Add R between dash numbers for recessed head.

—— AN520 ——

SCREW, ROUND HEAD (FINE THREAD)

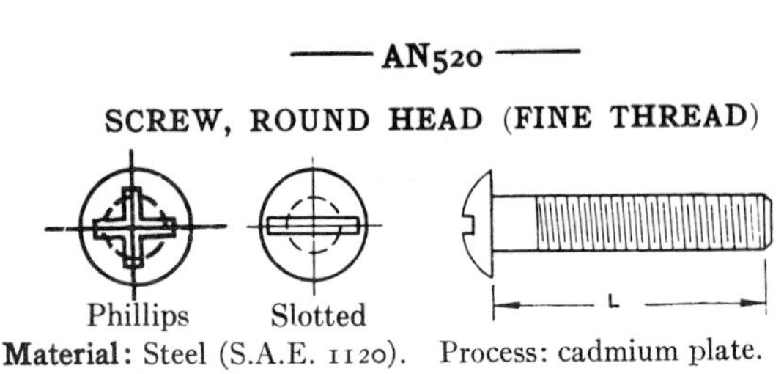

Phillips Slotted

Material: Steel (S.A.E. 1120). Process: cadmium plate.
Brass.
Aluminum alloy, "Inactive for design."
Corrosion-resistant steel, passivated.

Thread: Class 2 NF.

Sizes: No. 0 through 1/4 dia., coded same as AN515.

Example: AN520B10–16.

Add R between dash numbers for recessed head.

—— AN525 ——

SCREW, WASHER HEAD

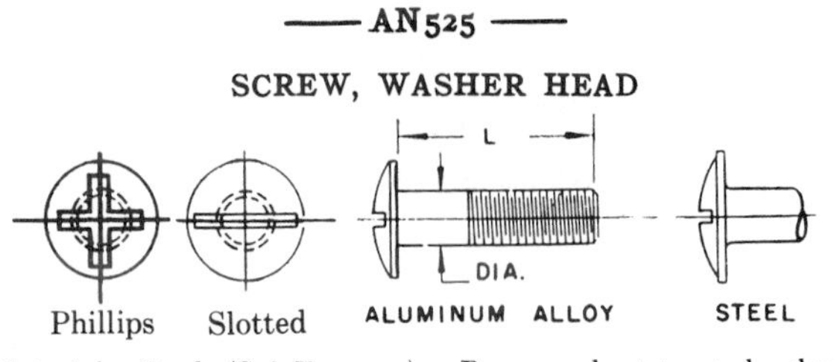

Phillips Slotted

Material: Steel (S.A.E. 2330). Process: heat-treat harden and cadmium plate.

Aluminum alloy. Process: heat-treat harden and anodize. D in code.

Thread: Class 3 NF.

Size: 8–36, 1032, and ½–28 coded according to Table 2, page 144 in first dash number. Length, given in second dash number, is in 16ths of an inch according to Table 3, page 144.

Examples: AN525-8-8 is washer-head screw, of steel, size 8–36, ½ in. long. AN525D8-8 is same screw made of aluminum alloy.

Add R between dash numbers for recessed head.

—— AN526 ——

SCREW, TRUSS HEAD

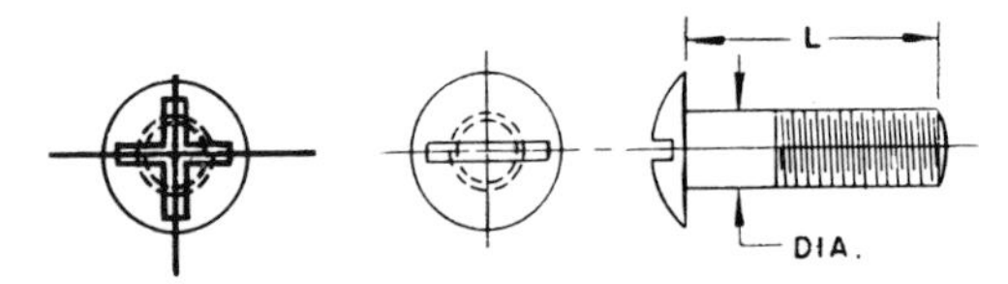

Material: Steel (low carbon—S.A.E. 1120). Process: cadmium plate.

Corrosion-resistant steel, passivated, indicated by C.

Aluminum alloy. Process: heat-treat harden and anodize. DD in code.

Thread: Class 2 NF and Class 2 NC.

Size: No. 6, No. 8, No. 10, and ¼ in. dia. indicated by first dash number according to Table 1, page 144. Length in 16ths of an inch given by second dash number according to Table 3, page 144.

Examples: AN526-640-10 is button-head screw, made of low-carbon steel, size 6–40, ⅝ in. long. AN526C640-10 is same screw made of corrosion-resistant steel. AN526DD640-10 is same screw made of aluminum alloy.

Add R between dash numbers for recessed head.

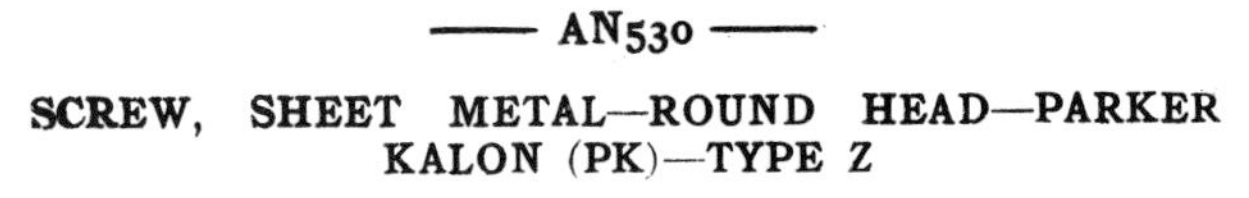

—— AN530 ——

SCREW, SHEET METAL—ROUND HEAD—PARKER KALON (PK)—TYPE Z

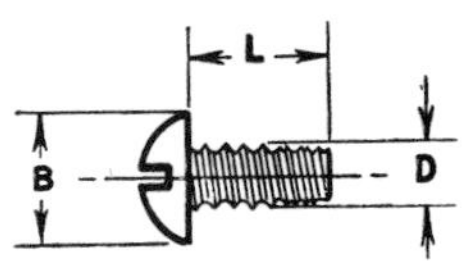

Material: Steel. Process: harden and cadmium plate (Spec. AN-QQ-P-421).

Description: Thread starts and makes its own threads in sheet metal owing to the reduced and grooved first two threads.

Size: First dash number gives screw size (−2, −4, −6, −8, −10, and −14) according to Table 2, page 144, of nominal diameter D, while second dash number gives L in 16ths of an inch.

Metal thickness	Hole sizes for screws					
	−2	−4	−6	−8	−10	−14 (¼ D.)
.015 to .028	.063	.086	.104	.116	.128	
.031 to .051	.073	.093	.110	.120	.136	.189
.063 to .081	.076	.101	.120	.140	.152	.201

Example: AN530–6–6 is sheet-metal screw, size 6 and ⅜ in. long.

—— AN531 ——

SCREW, SHEET METAL—FLAT HEAD—PARKER KALON (PK)—TYPE Z

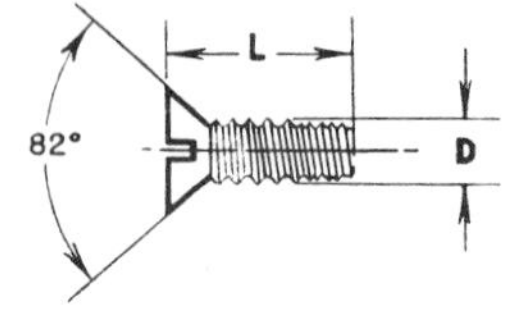

Material: Same as given with AN530, except that first dash numbers range from −4 to −14 only.

Example: AN531–6–6 is sheet-metal screw, size 6 and ⅜ in. long.

—— AN490 ——

ROD END, THREADED

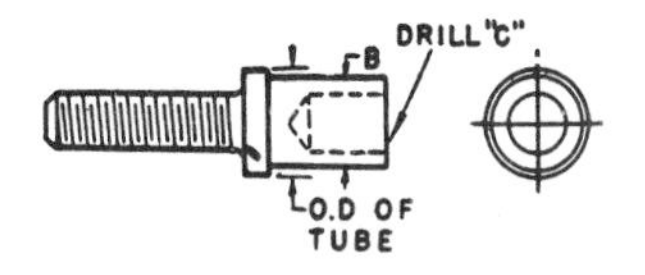

Material: Steel (Spec. AN–QQ–S–646). Finish: Cadmium plate indicated by P following dash number. Unfinished parts to be finished at assembly.

Thread: ¼–28 NF3 external.

SIZE

Dash No.	*A*, in.	*B*, in.	*C*, in.
−5	5/16	.2425	⅛
−6	⅜	.305	3/16
−8	½	.430	9/32

Examples: AN490–6 is threaded rod end for a ⅜ tube. AN490–6P is cadmium-plated.

AN666
MS21259
TERMINAL, THREADED CABLE (FOR SWAGING)

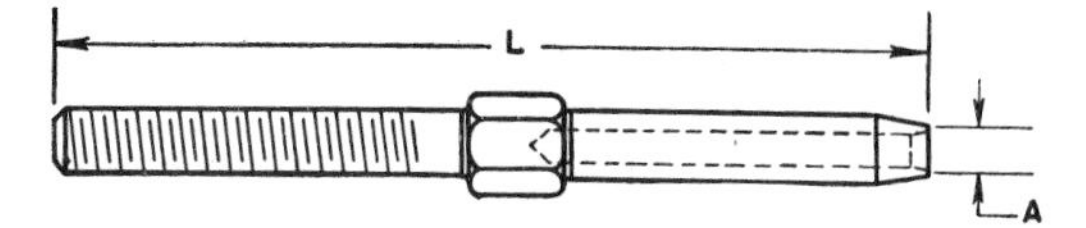

Sizes: See table under AN669.
Threads: Bolt R and L.
Example: AN666–8R is terminal, ¼ dia. cable, right-hand threads.

AN667
MS20667
TERMINAL, FORK END CABLE (FOR SWAGING)

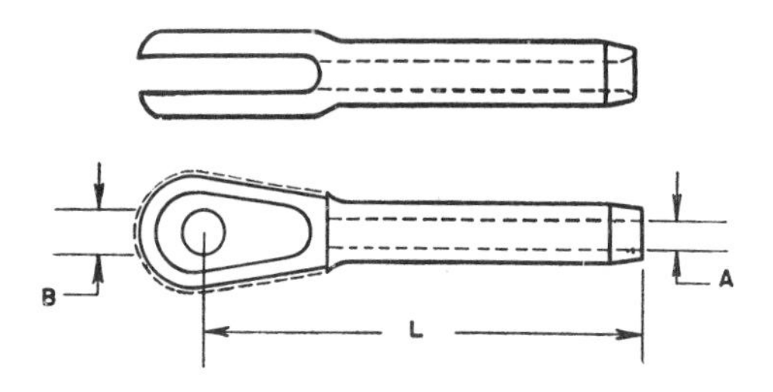

Sizes: See table under AN669.
Example: AN667–4 is terminal, ⅛ in. dia. cable, fork end.

AN668
MS20668
TERMINAL, EYE END CABLE (FOR SWAGING)

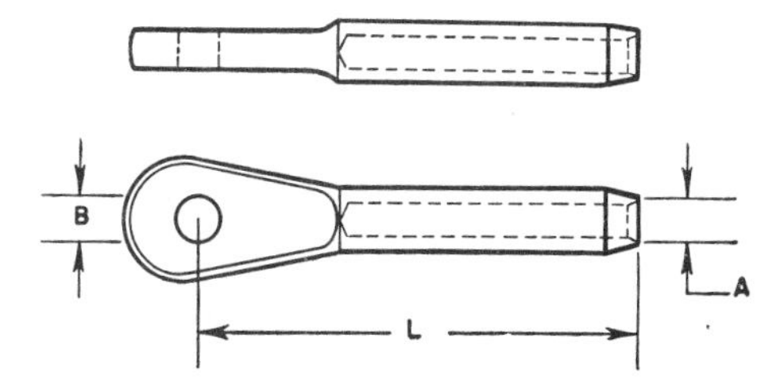

Sizes: See table under AN669.
Example: AN668–6 is terminal, 3/16 dia. cable, eye end.

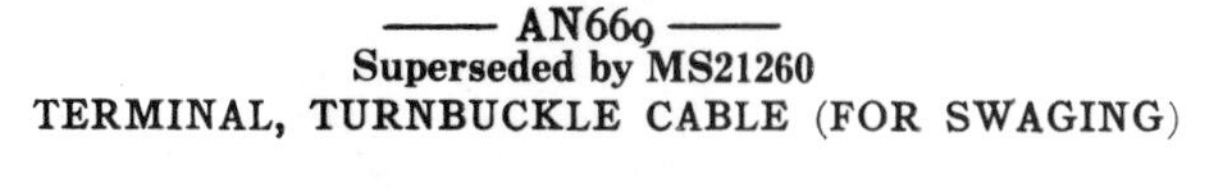

—— AN669 ——
Superseded by MS21260
TERMINAL, TURNBUCKLE CABLE (FOR SWAGING)

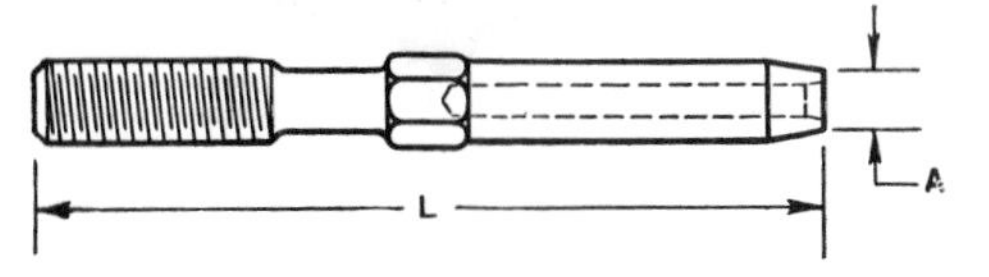

Sizes: See table. Dash numbers –3, –4, –5, –6 are both long and short sizes.

Threads: Both R and L.

Dash No.	Cable dia., in.	Thread NF-3 AN666 and AN669	A, in.	B, in.	Length before swaging, in.				
					AN666	AN667	AN668	AN669	AN669–S
–2	1/16	6–40	.078	.188	2.473	1.572	1.631	2.616	
–3	3/32	10–32	.109	.188	2.879	1.945	2.043	3.738	2.863
–4	1/8	1/4–28	.141	.188	3.333	2.352	2.337	4.020	3.145
–5	5/32	1/4–28	.172	.250	3.627	2.655	2.684	4.314	3.429
–6	3/16	5/16–24	.203	.313	4.002	3.071	3.019	4.612	3.737
–7	7/32	3/8–24	.234	.313	4.516	3.440	3.382	4.914	
–8	1/4	3/8–24	.265	.375	4.937	3.806	3.763	5.218	
–9	9/32	7/16–20	.297	.438	5.391	4.120	4.153	5.542	
–10	5/16	1/2–20	.328	.438	5.844	4.438	4.546	5.875	

Examples: AN669–L6LH is terminal, long, 3/16 dia. cable, left-hand threads. AN669–S6RH is terminal, short, 3/16 dia. cable, right-hand threads.

MS TURNBUCKLES (CLIP-LOCKING)

Clip-Locking Turnbuckles utilize two locking clips instead of lockwire for safetying. The turnbuckle barrel and terminals are slotted lengthwise to accommodate the locking clips. After the proper cable tension is reached the barrel slots are aligned with the terminal slots and the clips are inserted. The curved end of the locking clips expand and latch in the vertical slot in the center of the barrel.

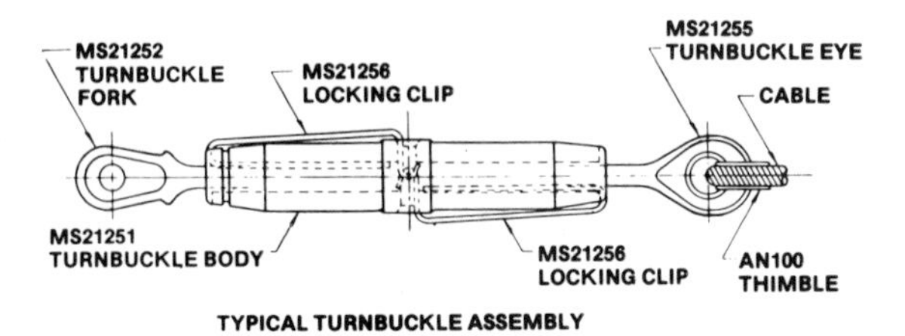

TYPICAL TURNBUCKLE ASSEMBLY

MS Standard Drawings for clip-locking turnbuckles supersede various AN Drawings for conventional (lockwire type) turnbuckle parts and NAS Drawings for clip-locking turnbuckle parts. Refer to the following cross reference tables for AN and NAS equivalents.

MS21251 TURNBUCKLE BARREL

Supersedes AN155 and NAS649 barrels. MS21251 items can replace AN155 items of like material and thread, but the AN155 items cannot replace the MS21251 items. MS21251 items are interchangeable with the NAS649 items of like material and thread. MS21251 barrels are available in brass (QQ-B-637, composition 2 or MIL-T-6945), steel (cadmium plated to QQ-P-416, type 2, class 3) or aluminum alloy (anodized to MIL-A-8725). The cross reference table shows equivalent items made of brass.

MS21251 DASH NO.	ROPE DIA.	THREAD SIZE	AN155 DASH NO	NAS649 DASH NO.	USES MS21256 CLIP DASH NO.
B2S	1/16	6-40	B8S	B8S	-1
B2L	1/16	6-40	B8L	B8L	-2
B3S	3/32	10-32	B16S	B16S	-1
B3L	3/32	10-32	B16L	B16L	-2
B5S	5/32	1/4-28	B32S	B32S	-1
B5L	5/32	1/4-28	B32L	B32L	-2
B6S	3/16	5/16-24	B46S	B46S	-1
B6L	3/16	5/16-24	B46L	B46L	-2
B8L	1/4	3/8-24	B80L	B80L	-2
B9L	9/32	7/16-20	B125L	B125L	-3
B10L	5/16	1/2-20	B175L	B175L	-3

TERMINALS

MS items can replace AN items of like thread except for the -22 and -61 sizes, but the AN items cannot replace the MS items. MS items are interchangeable with the NAS items of like thread except for the -22 and -61 sizes. These MS terminals are available only in steel cadmium plated to QQ-P-416, type 2, class 3. Available with right-hand (R) or left-hand (L) threads.

MS21252 TURNBUCKLE FORK supersedes AN161 and NAS645 forks.

MS21254 PIN EYE supersedes AN165 and NAS648 eyes.

MS21255 CABLE EYE supersedes AN170 and NAS 647 eyes.

MS21260 SWAGED STUD END supersedes AN669 studs.

MS21252 MS21254 MS21255 DASH NOS.		WIRE ROPE DIA.	THREAD SIZE	AN 161 AN165 AN170 DASH NOS.		NAS645 NAS648 NAS647 DASH NOS.	
RH THD	LH THD			RH THD	LH THD	RH THD	LH THD
-2RS	-2LS	1/16	6-40	-8RS	-8LS	-8RS	-8LS
-2RL*	-2LL*	1/16	6-40	—	—	—	—
-3RS	-3LS	3/32	10-32	-16RS	-16LS	-16RS	-16LS
-3RL	-3LL	3/32	10-32	-16RL	-16LL	-16RL	-16LL
-5RS	-5LS	5/32	1/4-28	-32RS	-32LS	-32RS	-32LS
-5RL	-5LL	5/32	1/4-28	-32RL	-32LL	-32RL	-32LL
-6RS	-6LS	3/16	5/16-24	-46RS	-46LS	-46RS	-46LS
-6RL	-6LL	3/16	5/16-24	-46RL	-46LL	-46RL	-46LL
-8RL	-8LL	1/4	3/8-24	-80RL	-80LL	-80RL	-80LL
-9RL	-9LL	9/32	7/16-20	-125RL	-125LL	-125RL	-125LL
-10RL	-10LL	5/16	1/2-20	-175RL	-175LL	-175RL	-175LL

*MS21254 and MS21255 eyes only; MS21252 fork not made in this size.

MS21256 TURNBUCKLE CLIP

Made of corrosion resistant steel wire, QQ-W-423, composition FS302, condition B. These are NOT interchangeable with the NAS651 clips. Available in 3 sizes: MS21256-1, -2 and -3. For applications, see the MS21251 Turnbuckle Barrel Cross Reference Chart.

PART NUMBER	THREAD	CABLE DIA.	DESCRIPTION
MS21251-B2S	6-40	1/16	Barrel (Body), Brass
-B3S	10-32	3/32	
-B3L	10-32	3/32	
-B5S	¼-28	5/32	
-B5L	¼-28	5/32	
MS21252-3LS	10-32	3/32	Fork (Clevis End)
-3RS	10-32	3/32	
-5RS	¼-28	5/32	
MS21254-2RS	6-40	1/16	Eye End (for pin)
-3LS	10-32	3/32	
-3RS	10-32	3/32	
-5LS	¼-28	5/32	
-5RS	¼-28	5/32	

PART NUMBER	THREAD	CABLE DIA.	DESCRIPTION
MS21255-3LS	10-32	3/32	Eye End (for cable)
-3RS	10-32	3/32	
MS21256-1	—	—	Clip (for short barrels)
-2	—	—	Clip (for long barrels)
MS21260-S2LH	6-40	1/16	End (for cable)
-S2RH	6-40	1/16	
-S3LH	10-32	3/32	
-S3RH	10-32	3/32	
-L3LH	10-32	3/32	
-L3RH	10-32	3/32	
-S4LH	¼-28	1/8	
-S4RH	¼-28	1/8	
-L4LH	¼-28	1/8	
-L4RH	¼-28	1/8	

MS21260 SWAGED STUD END

These clip-locking terminals are available in corrosion resistant steel and in cadmium plated carbon steel. MS21260 items can replace AN669 items of the same dash numbers, but the AN669 items cannot always replace the MS21260 items.

Example: The AN "equivalent" (the AN equivalent would not be clip-locking) for MS21260 L3RH would be AN669-L3RH. There would be no AN equivalent for a MS21260FL3RH, since AN669 terminals are not available in carbon steel.

—— AN774 to AN932 ——
PLUMBING FITTINGS

Material:

Aluminum alloy .... (code D)
Steel .... (code, absence of letter)
Brass .... (code B)
Aluminum bronze .... (code Z—for AN819 sleeve)

Size: The dash number following the AN number indicates the size of the tubing (or hose) for which the fitting is made, in 16ths of an inch. This size measures the O. D. of tubing and the I. D. of hose. Fittings having pipe threads are coded by a dash number, indicating the pipe size in 8ths of an inch. The material code letter, as noted above, follows the dash number.

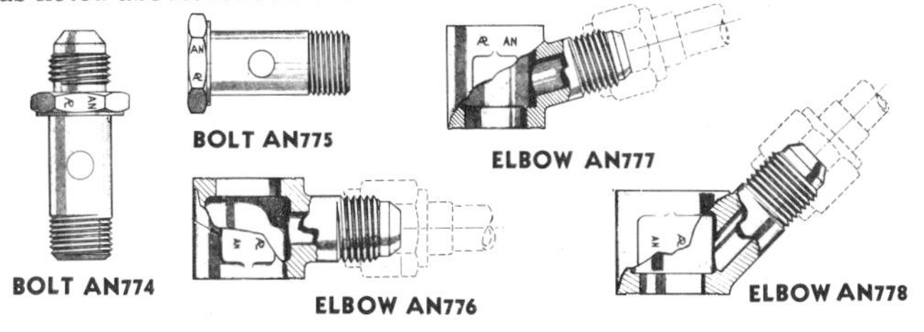

BOLT AN775

ELBOW AN777

BOLT AN774

ELBOW AN776

ELBOW AN778

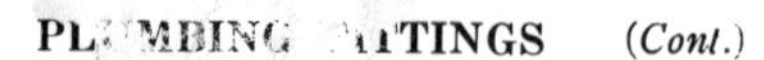

PLUMBING FITTINGS (*Cont.*)

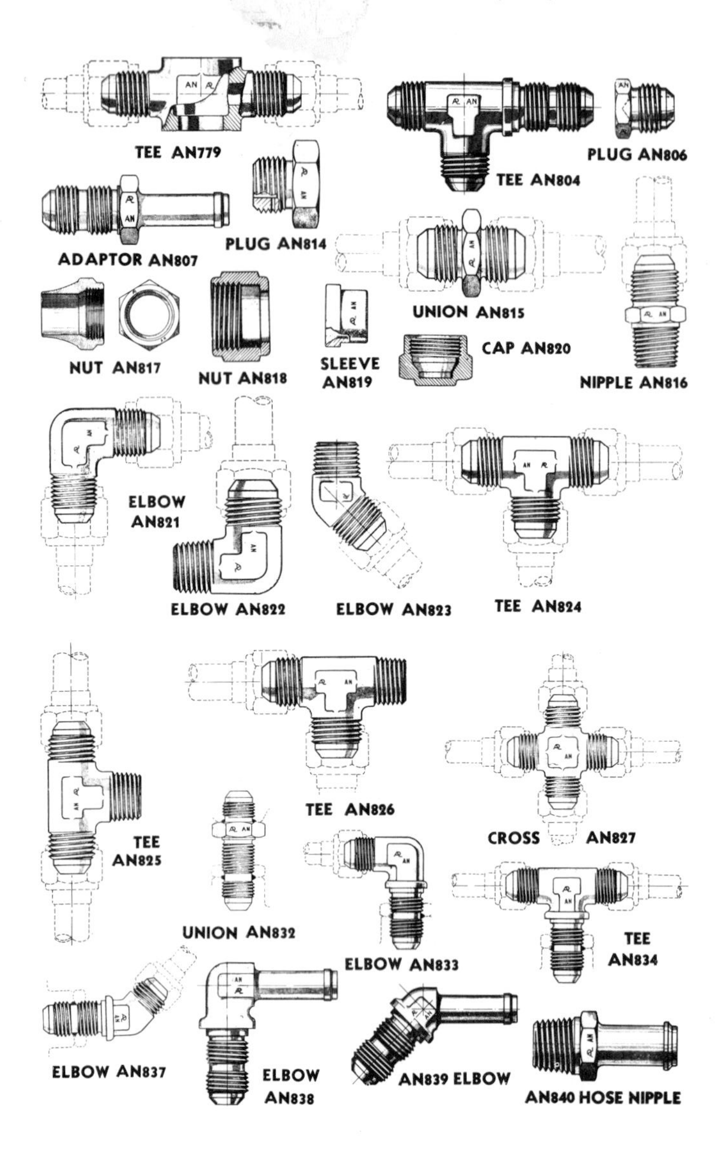

PLUMBING FITTINGS (Cont.)

★ Inactive for new design.

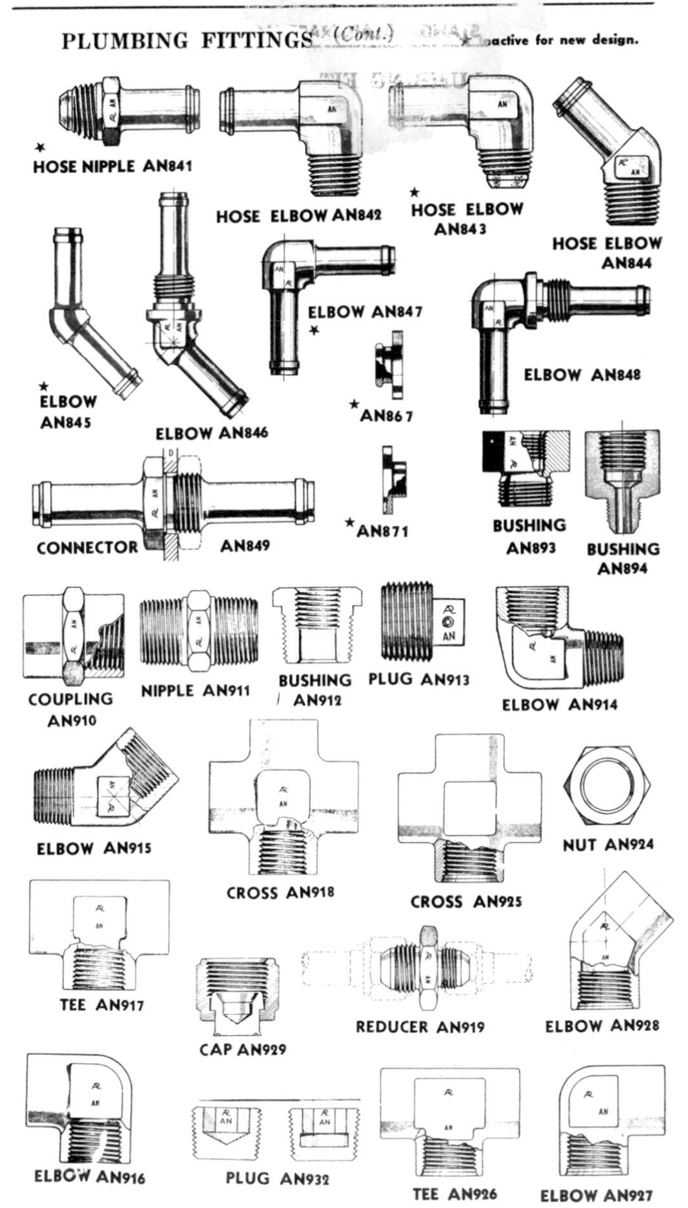

— AN809 —

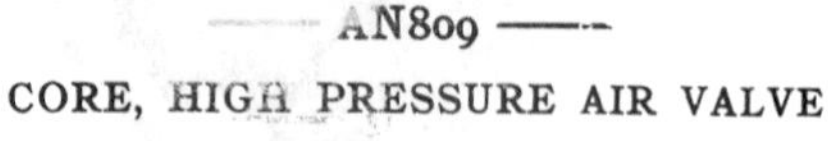

CORE, HIGH PRESSURE AIR VALVE

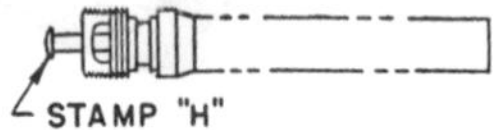

Symbol H stamped on head for pressure identification.
Size: AN809–1 is to be used with AN812–1 body.

— AN811 —

FITTINGS, SOLDERLESS

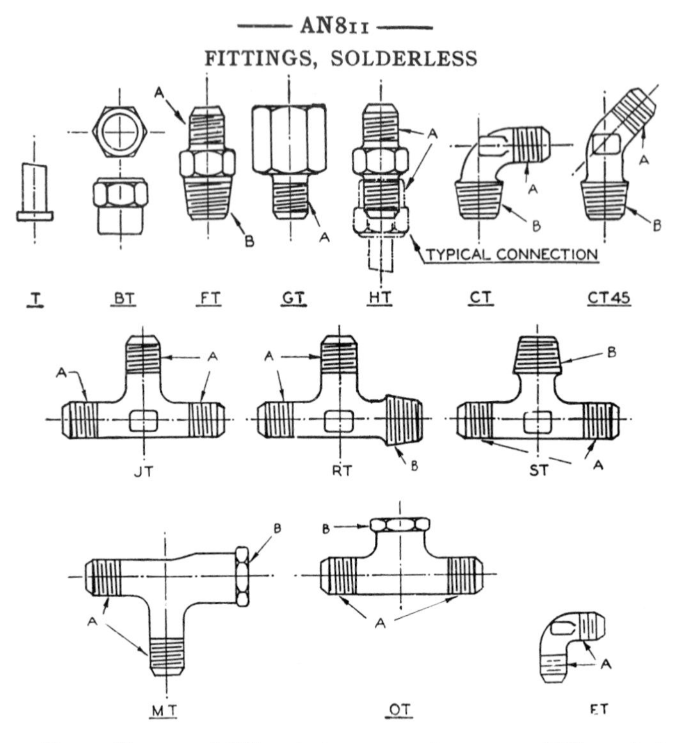

NOTE: Sleeve and BT nut shown as reference on (HT) typical assembly.

Material: Sleeves T shall be copper-silicon alloy or nickel steel, heat-treated.

Nuts and fittings shall be brass, aluminum alloy, or nickel steel, heat-treated.

Brass, designated by absence of letter following dash number.

Aluminum alloy, designated by D following dash number.

Copper silicon, designated by CS following dash number.

AN931

GROMMET, NEOPRENE

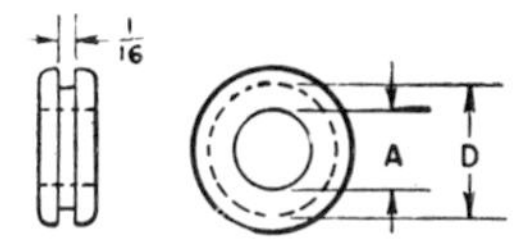

Material: Neoprene, durometer hardness 50–55.

Sizes: First dash number is size of hole, *A* dimension (see Fig.) in 16ths of an inch. Second dash number is *D* dimension in 16ths. The dash numbers are –2–16; –3–5; –3–9; –4–7; –4–16; –5–9; –5–12; –6–10; –6–16; –7–11; –8–13; –8–20; –9–13; –10–14; –10–20; –11–16; –12–17; –12–20; –12–23; –14–20; –14–26; –16–22; –16–30; –20–38; –24–28.

Example: AN931–6–10 is Neoprene grommet, 3/8 I.D., 5/8 O.D.

AN935

WASHER, LOCK

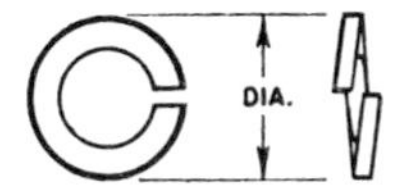

Material: Steel, Spec. 25523.
Phospherous bronze.

Sizes: Washers are made in regular and light series. Light series are of narrower and thinner stock; L in code. Dash numbers given in the table.

Dash No.	Bolt size	Dash No.	Bolt size	Dash No.	Bolt size
–2	No. 2(.086)	–12	No. 12(.216)	–816	1/2
–4	No. 4(.112)	–416	1/4	–916	9/16
–6	No. 6(.138)	–516	5/16	–1016	5/8
–8	No. 8(.164)	–616	3/8	–1216	3/4
–10	No. 10(.190)	–716	7/16		

Examples: AN935–10 is lock washer for No. 10 bolt, regular series. AN935–10L indicates light series. Add B for bronze.

—— AN936 ——

WASHER, LOCK—SHAKEPROOF

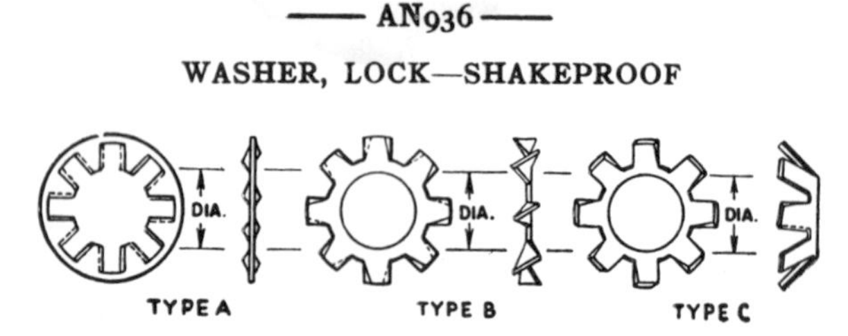

Material: Steel or bronze.

Finish: Steel. Process: cadmium plate, Spec. AN–QQ–P–421. Bronze. Process: tinned.

Sizes: Letters A, B, or C before dash number indicate type (see drawing). Letter B after dash number indicates bronze.

Type A: Dash numbers are –2 to –10 and –416 to –716 incl. (Refer to table under AN935 for bolt sizes.)

Type B: Dash numbers are –4 to –10 and –416 to –716 incl.

Type C: Dash numbers are –6 to –10 and –416 to –716 incl.

Examples: AN936A416B is washer, type A, for ¼ bolt, bronze. AN936B416 is type B, steel. AN936C416 is type C, steel.

—— AC940 ——

OBSOLETE

WASHER, BURR

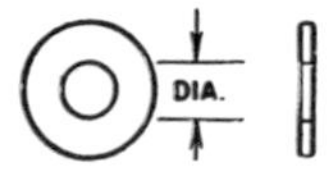

Material: Tinned steel, brass, copper, aluminum, or aluminum alloy.

Sizes: Taken from United Screw & Bolt Corp. catalogue.

Dash No.	Screw, bolt or rivet size	Dash No.	Screw, bolt or rivet size	Dash No.	Screw, bolt or rivet size
–14	1⁄16(.083 dia.)	–9	No. 6(.148 dia.)	–5	.220 dia.
–13	3⁄32 dia.	–8	No. 8(5⁄32 dia.)	–416	¼ dia.
–12	No. 4(.109)	–7	.180 dia.	–516	5⁄16 dia.
–11	No. 5(⅛)	–316	No. 10(3⁄16 dia.)	–616	⅜ dia.

Examples: AC940–14 is washer, steel, for 1⁄16 rivet. AC940B14 indicates brass. AC940C14 indicates copper. AC940A14 indicates aluminum. AC940D14 indicates aluminum alloy.

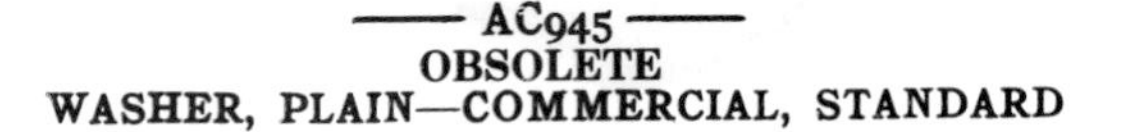

—— AC945 ——
OBSOLETE
WASHER, PLAIN—COMMERCIAL, STANDARD

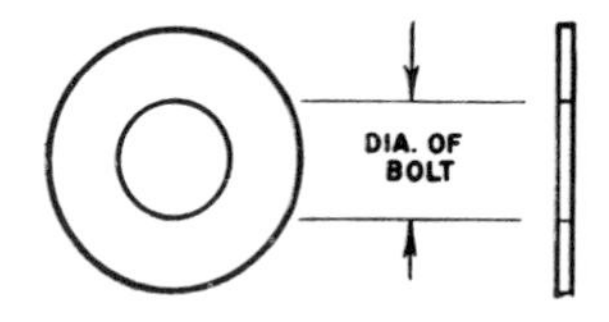

Material: Wrought iron or steel.

Sizes: Dash numbers are –4 to –24 incl. They indicate bolt size in 16ths of an inch.

Example: AC945–18 is washer, wrought iron or steel, for 1⅛ bolt.

—— AN960 ——

WASHER, PLAIN

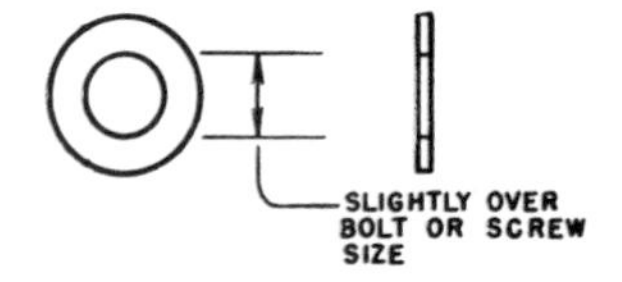

Materials: Carbon steel, Spec. AN–QQ–S–651 or S.A.E.–1010 hard finished.

Corrosion-resistant (stainless) steel, Spec. AN–QQ–S–757; C in code.

Aluminum, Spec. QQ–A–561; A in code.

Aluminum alloy, Spec. QQ–A–353; D in code.

Brass, Spec. QQ–B–611; B in code.

Carbon steel, aluminum, and stainless are made in both regular and light series. Light series are made of thinner stock and are indicated by L following the dash number.

Finish: Carbon steel. Process: cadmium plate, Spec. AN–QQ–P–421.

Aluminum alloy. Process: anodize, Spec. AN–QQ–A–696.

Sizes: Dash numbers –3, –4, –6, –8, and –10 are machine screw sizes. –416 to –4016 are bolt sizes in 16ths.

Examples: AN960–4 is carbon steel washer for No. 4 screw. AN960C416 is stainless steel washer for ¼ bolt. AN960A4016L is aluminum washer, 2½-in. bolt, light series. AN960B1716 is brass washer, 1$\frac{1}{16}$ bolt. AN960D1716 indicates aluminum alloy.

WASHER, FLAT—FOR WOOD

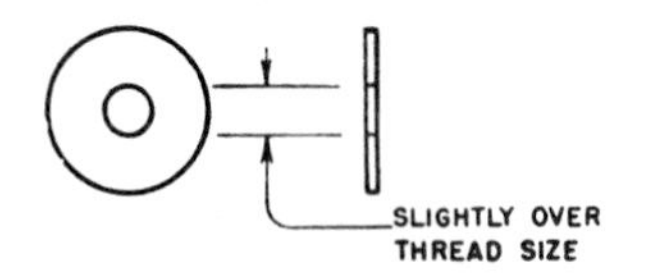

Material: Steel. Finish: cadmium plate, Spec. AN–QQ–P–421.
Sizes: Dash numbers –3 to –10 incl. are bolt sizes in 16ths.
Example: AN970–6 is washer for ⅜ bolt.

WASHER, TAPER PIN

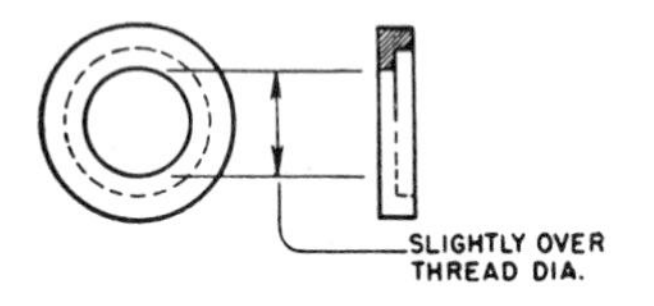

Used with AC386 taper pins.

Material: Steel, Spec. AN–QQ–S–646 or AN–QQ–S–651. Finish: cadmium plate, Spec. AN–QQ–P–421.

Sizes: Dash numbers –3 to –14 incl. are bolt sizes in 16ths.

Example: AC975–5 is washer for taper pin with 5⁄16 bolt thread.

AN995
Superseded by MS20995
WIRE, FOR LOCKING

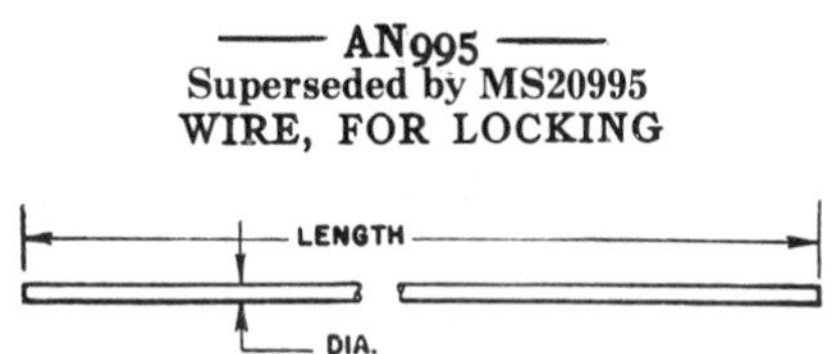

Material: Copper, Spec. 57–222–1; C in code.
Galvanized steel, Spec. 48–19.

Sizes: Wire diameters are .032, .040, .051, .072, and .090, indicated by first dash number. Standard lengths are 1, 2, 3, 4, 6, 8, 10, 14, 18, 24, 36, 48, 60, 72, 84, and 100 in., indicated by second dash number.

—NAS1—

Superseded by MS20002

WASHER, HIGH STRENGTH BOLT

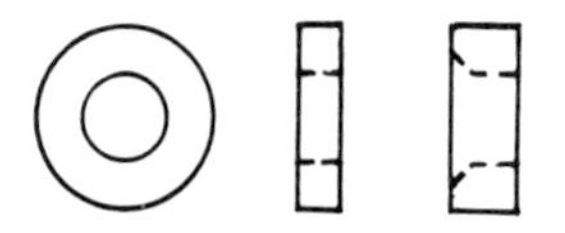

Material: Steel.

Code: A dash number is added to the part number, indicating the diameter of the bolt it fits in 16ths of an inch.

—NAS144 to NAS158—

BOLT-INTERNAL HEX HEAD

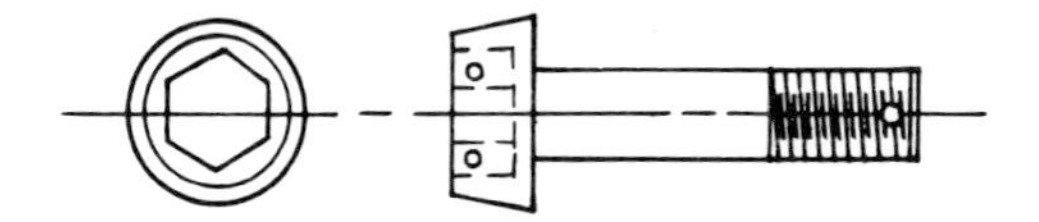

Material: Steel, heat treated to 160,000 - 180,000 p.s.i. Finish: Cadmium plate.

Code: Diameter, from 1/4 inch to 1-1/8 inch, indicated by the last digit of the NAS number in 16ths of an inch.

Length, to 8 inches, is indicated by the dash number in 16ths of an inch. (-16 is one inch long; -32 is two inches long, and -38 is 2-3/8 inches long).

Add D. H. after the part number to indicate drilled head.

—NAS334 to NAS340—

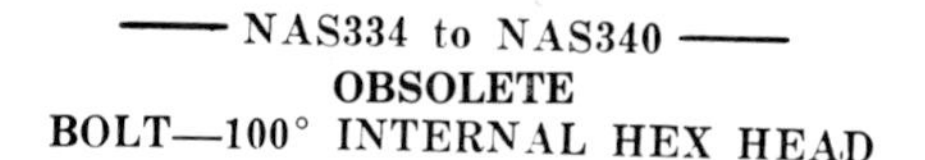

OBSOLETE

BOLT—100° INTERNAL HEX HEAD

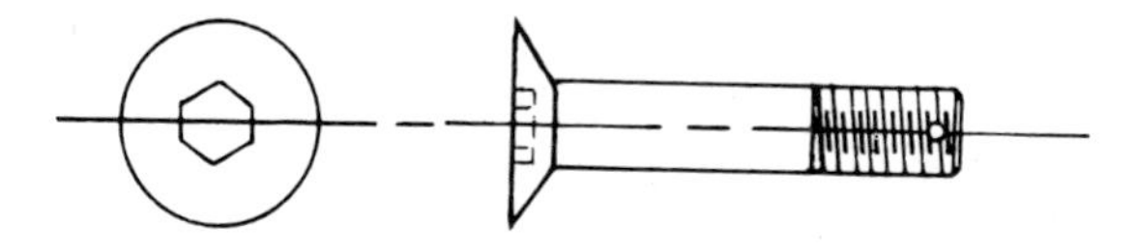

Material and **Code** like NAS144 to NAS158

TABLE 1. DASH NO. (SIZE) CHART, SCREW PARTS

FINE THREAD (NF)

Size and thread	Dash No.	Dia. and thread	Dash No.	Dia. and thread	Dash No.	Dia. and thread	Dash No.
4–48	–448	1/4–28	–428	1/2–20	–820	7/8–14	–1414
6–40	–640	5/16–24	–524	9/16–18	–918	1–14	–1614
8–36	–836	3/8–24	–624	5/8–18	–1018		
10–32	–1032	7/16–20	–720	3/4–16	–1216		

COARSE THREAD (NC)

Size and thread	Dash No.	Size and thread	Dash No.	Size and thread	Dash No.
4–40	–440	10–24	–1024	3/8–16	–616
6–32	–632	1/4–20	–420	7/16–14	–714
8–32	–832	5/16–18	–518	1/2–13	–813

TABLE 2. DASH NO. (SIZE) CHART—PARTS MADE IN ONE THREAD ONLY (NF OR NC)

Size	Dash No.	Size	Dash No.	Size (dia.), in.	Dash No.	Dia., in.	Dash No.
0	–0	5	–5	1/4	–416	1/2	–816
1	–1	6	–6	5/16	–516	9/16	–916
2	–2	8	–8	3/8	–616	5/8	–1016
3	–3	10	–10	7/16	–716		
4	–4						

TABLE 3. DASH NO. (LENGTH) CHART—PART LENGTHS IN 16THS OF AN INCH

Length, in.	Dash No.	Length, in.	Dash No.	Length, in.	Dash No.	Length, in.	Dash No.
1/4	–4	5/8	–10	1 3/8	–22	2 1/4	–36
5/16	–5	3/4	–12	1 1/2	–24	2 1/2	–40
3/8	–6	7/8	–14	1 5/8	–26	2 3/4	–44
7/16	–7	1	–16	1 3/4	–28	3	–48
1/2	–8	1 1/8	–18	1 7/8	–30		
9/16	–9	1 1/4	–20	2	–32		

TABLE 4. DASH NO. (SIZE) CHART—RIVETS

DIAMETER—FIRST DASH NO.

Dia., in.	Dash No.	Dia., in.	Dash No.	Dia., in.	Dash No.	Dia., in.	Dash No.	Dia., in.	Dash No.
1/16	–2	1/8	–4	3/16	–6	5/16	–10	7/16	–14
3/32	–3	5/32	–5	1/4	–8	3/8	–12		

LENGTH—SECOND DASH NO.

L, in.	Dash No.	*L*, in.	Dash No.	*L*, in.	Dash No.	*L*, in.	Dash No.
3/16	–3	9/16	–9	1 1/8	–18	2	–32
1/4	–4	5/8	–10	1 1/4	–20	2 1/2	–40
5/16	–5	3/4	–12	1 3/8	–22	3	–48
3/8	–6	7/8	–14	1 1/2	–24	3 1/2	–56
7/16	–7	1	–16	1 3/4	–28	4	–64
1/2	–8						

CHAPTER IX

REFERENCE TABLES

AIRCRAFT THREAD AND DRILL SIZES

The screw and nut combination is probably the most used of machine elements. Certain thread series have, through their extensive use, come to be known as "standard."

Classification of Threads

Aircraft bolts, screws, and nuts are threaded in either the NC (American National Coarse) thread series, the NF (American National Fine) thread series, the UNC (American Standard Unified Coarse) thread series, or the UNF (American Standard Unified Fine) thread series. Although they are interchangeable there is one difference

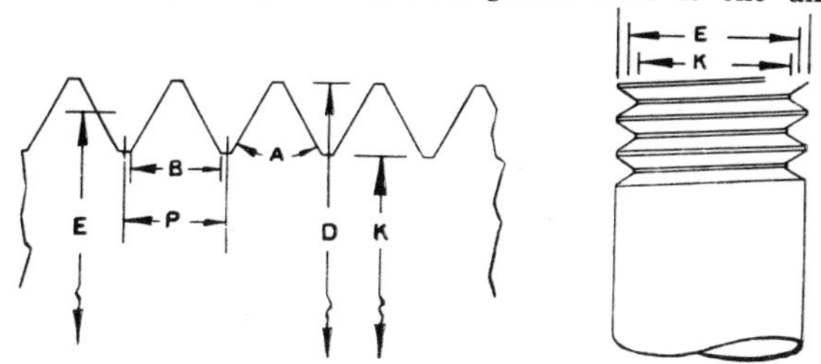

FIG. 1.—A, thread angle; B, base; D, major diameter; E, pitch diameter; H, depth; K, minor diameter; P, pitch.

between the American National series and the American Standard Unified series that should be pointed out. In the 1-inch diameter size, the NF thread specified 14 threads per inch (1-14NF), while the UNF thread specifies 12 threads per inch (1-12UNF). Both type threads are designated by the number of times the incline (threads) rotates around a 1-inch length of a given diameter bolt or screw. For example, a 4-28 thread indicates that a ¼-inch diameter bolt has 28 threads in 1 inch of its threaded length.

Threads are also designated by Class of fit. The Class of a thread indicates the tolerance allowed in manufacturing. Class 1 is a loose fit, Class 2 is a free fit, Class 3 is a medium fit, and Class 4 is a close fit. **Aircraft bolts are almost always manufactured in the Class 3, medium fit.** A Class 4 fit requires a wrench to turn the nut onto a bolt whereas a Class 1 fit can easily be turned with the fingers. Generally aircraft screws are manufactured with a Class 2 thread fit for ease of assembly. The general purpose aircraft bolt, AN3 thru AN20 has UNF-3 threads (American Standard Unified Fine, Class 3, medium fit).

Bolts and nuts are also produced with right-hand and left-hand threads. A right-hand thread tightens when turned clockwise; a left-hand thread tightens when turned counterclockwise. Except in special cases, all aircraft bolts and nuts have right hand threads.

TAP DRILL SIZES

National Coarse Thread Series Medium Fit, Class 3 (NC)

Size and threads	Dia. of body	Body drill	Preferred dia. of hole	Tap drill
1–64	.073	47	.0575	No. 53
2–56	.086	42	.0682	No. 51
3–48	.099	37	.078	5/64 in.
4–40	.112	31	.0866	No. 44
5–40	.125	29	.0995	No. 39
6–32	.138	27	.1063	No. 36
8–32	.164	18	.1324	No. 29
10–24	.190	10	.1476	No. 26
12–24	.216	2	.1732	No. 17
1/4–20	.250	1/4	.1990	No. 8
5/16–18	.3125	5/16	.2559	*F*
3/8–16	.375	3/8	.3110	5/16 in.
7/16–14	.4375	7/16	.3642	*U*
1/2–13	.500	1/2	.4219	27/64 in.
9/16–12	.5625	9/16	.4776	31/64
5/8–11	.625	5/8	.5315	17/32 in.
3/4–10	.750	3/4	.6480	41/64 in.
7/8–9	.875	7/8	.7307	49/64 in.
1–8	1.000	1	.8376	7/8 in.

National Fine Thread Series Medium Fit, Class 3 (NF)

Size and threads	Dia. of body	Body drill	Preferred dia. of hole	Tap drill
0–80	.060	52	.0472	3/64 in.
1–72	.073	47	.0591	No. 53
2–64	.086	42	.0700	No. 50
3–56	.099	37	.0810	No. 46
4–48	.112	31	.0911	No. 42
5–44	.125	29	.1024	No. 38
6–40	.138	27	.113	No. 33
8–36	.164	18	.136	No. 29
10–32	.190	10	.159	No. 21
12–28	.216	2	.180	No. 15
1/4–28	.250	*F*	.213	No. 3
5/16–24	.3125	5/16	.2703	*I*
3/8–24	.375	3/8	.332	*Q*
7/16–20	.4375	7/16	.386	*W*
1/2–20	.500	1/2	.449	7/16 in.
9/16–18	.5625	9/16	.506	1/2 in.
5/8–18	.625	5/8	.568	9/16 in.
3/4–16	.750	3/4	.6688	11/16 in.
7/8–14	.875	7/8	.7822	51/64 in.
1–14	1.000	1	.9072	59/64 in.

National Taper Pipe Thread

Size pipe thread, in.	No. of threads per inch	Outside dia. of pipe for threading		Size pipe reamer, in.	Size tap drill, in.
		Decimal inch	Nearest fraction of inch		
1/8	27	.405	13/32	1/8	21/64
1/4	18	.540	35/64	1/4	7/16
3/8	18	.675	43/64	3/8	9/16
1/2	14	.840	27/32	1/2	45/64
3/4	14	1.050	13/64	3/4	29/32

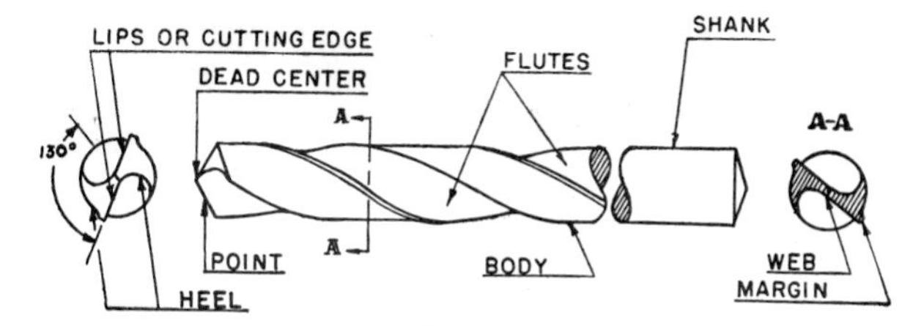

FIG. 2.

TWIST DRILL SIZES

Decimal equivalent	Fraction	Number or letter	Decimal equivalent	Fraction	Number or letter	Decimal equivalent	Fraction	Number or letter
.0135	...	80	.096*	....	41	.2187	7/32	
.0145	...	79	.098	....	40	.221	....	2
.0156	...	78	.0995	....	39	.228	....	1
.016	1/64		.1015	....	38	.234	....	A
.018	...	77	.104	....	37	.2343	15/64	
.020	...	76	.1065	....	36	.238	....	B
.021	...	75	.1093	7/64		.242	....	C
.0225	...	74	.110	....	35	.246	....	D
.024	...	73	.111	....	34	.250	1/4	E
.025	...	72	.113	....	33	.257	....	F
.026	...	71	.116	....	32	.261	....	G
.028	...	70	.120	....	31	.2656	17/64	
.029	...	69	.125	1/8		.266	....	H
.031	...	68	.1285	....	30	.272	....	I
.0313	...	67	.136	....	29	.277	....	J
.032	1/32		.1405	....	28	.281	....	K
.033	...	66	.1406	9/64		.2812	9/32	
.035	...	65	.144	....	27	.290	....	L
.036	...	64	.147	....	26	.295	....	M
.037	...	63	.1495	....	25	.2968	19/64	
.038	...	62	.152	....	24	.302	....	N
.039	...	61	.154	....	23	.3125	5/16	
.040	...	60	.1562	5/32		.316	....	O
.041	...	59	.157	....	22	.323	....	P
.042	...	58	.159	....	21	.3281	21/64	
.043	...	57	.161	....	20	.332	.....	Q
.0465	...	56	.166	....	19	.339	....	R
.0468	3/64		.1695	....	18	.3437	11/32	
.052	...	55	.1718	11/64		.348	....	S
.055	...	54	.173	....	17	.358	....	T
.0595	...	53	.177	....	16	.3594	23/64	
.0625	1/16		.180	....	15	.368	....	U
.0635	...	52	.182	....	14	.375	3/8	
.067	...	51	.185	....	13	.377	....	V
.070	...	50	.1875	3/16		.386	....	W
.073	...	49	.189	....	12	.3906	25/64	
.076	...	48	.191	....	11	.397	....	X
.0781	5/64		.1935	....	10	.404	....	Y
.0785	...	47	.196	....	9	.4062	13/32	
.081	...	46	.199	....	8	.413	....	Z
.082	...	45	.201	....	7	.4218	27/64	
.086	...	44	.2031	13/64		.4375	7/16	
.089	...	43	.204	....	6	.4531	29/64	
.0935	...	42	.2055	....	5	.4687	15/32	
.0937	3/32		.209	....	4	.4843	31/64	
			.213	....	3	.500	1/2	

FUNCTIONS OF NUMBERS

No.	Square	Cube	Square root	Cube root	Reciprocal	Circumference	Area
1	1	1	1.0000	1.0000	1.000000000	3.1416	0.7854
2	4	8	1.4142	1.2599	.500000000	6.2832	3.1416
3	9	27	1.7321	1.4422	.333333333	9.4248	7.0686
4	16	64	2.0000	1.5874	.250000000	12.5664	12.5664
5	25	125	2.2361	1.7100	.200000000	15.7080	19.635
6	36	216	2.4495	1.8171	.166666667	18.850	28.274
7	49	343	2.6458	1.9129	.142857143	21.991	38.485
8	64	512	2.8284	2.0000	.125000000	25.133	50.266
9	81	729	3.0000	2.0801	.111111111	28.274	63.617
10	100	1,000	3.1623	2.1544	.100000000	31.416	78.540
11	121	1,331	3.3166	2.2240	.090909091	34.558	95.033
12	144	1,728	3.4641	2.2894	.083333333	37.699	113.10
13	169	2,197	3.6056	2.3513	.076923077	40.841	132.73
14	196	2,744	3.7417	2.4101	.071428571	43.982	153.94
15	225	3,375	3.8730	2.4662	.066666667	47.124	176.71
16	256	4,096	4.0000	2.5198	.062500000	50.265	201.06
17	289	4,913	4.1231	2.5713	.058823529	53.407	226.98
18	324	5,832	4.2426	2.6207	.055555556	56.549	254.47
19	361	6,859	4.3589	2.6684	.052631579	59.690	283.53
20	400	8,000	4.4721	2.7144	.050000000	62.832	314.16
21	441	9,261	4.5826	2.7589	.047619048	65.973	346.36
22	484	10,648	4.6904	2.8020	.045454545	69.115	380.13
23	529	12,167	4.7958	2.8439	.043478261	72.257	415.48
24	576	13,824	4.8990	2.8845	.041666667	75.398	452.39
25	625	15,625	5.0000	2.9240	.040000000	78.540	490.87
26	676	17,576	5.0990	2.9625	.038461538	81.681	530.93
27	729	19,683	5.1962	3.0000	.037037037	84.823	572.56
28	784	21,952	5.2915	3.0366	.035714286	87.965	615.75
29	841	24,389	5.3852	3.0723	.034482759	91.106	660.52
30	900	27,000	5.4772	3.1072	.033333333	94.248	706.86
31	961	29,791	5.5678	3.1414	.032258065	97.389	754.77
32	1,024	32,768	5.6569	3.1748	.031250000	100.53	804.25
33	1,089	35,937	5.7446	3.2075	.030303030	103.67	855.30
34	1,156	39,304	5.8310	3.2396	.029411765	106.81	907.92
35	1,225	42,875	5.9161	3.2717	.028571429	109.96	962.11
36	1,296	46,656	6.0000	3.3019	.027777778	113.10	1,017.88
37	1,369	50,653	6.0828	3.3322	.027027027	116.24	1,075.21
38	1,444	54,872	6.1644	3.3620	.026315789	119.38	1,134.11
39	1,521	59,319	6.2450	3.3912	.025641026	122.52	1,194.59
40	1,600	64,000	6.3246	3.4200	.025000000	125.66	1,256.64
41	1,681	68,921	6.4031	3.4482	.024390244	128.81	1,320.25
42	1,764	74,088	6.4807	3.4760	.023809524	131.95	1,385.44
43	1,849	79,507	6.5574	3.5034	.023255814	135.09	1,452.20
44	1,936	85,184	6.6332	3.5303	.022727273	138.23	1,520.53
45	2,025	91,125	6.7082	3.5569	.022222222	141.37	1,590.43
46	2,116	97,336	6.7823	3.5830	.021739130	144.51	1,661.90
47	2,209	103,823	6.8557	3.6088	.021276600	147.65	1,734.94
48	2,304	110,592	6.9282	3.6342	.020833333	150.80	1,809.56
49	2,401	117,649	7.0000	3.6593	.020408163	153.94	1,885.74
50	2,500	125,000	7.0711	3.6840	.020000000	157.08	1,963.50

FUNCTIONS OF NUMBERS

No.	Square	Cube	Square root	Cube root	Reciprocal	Circumference	Area
51	2,601	132,651	7.1414	3.7084	.019607843	160.22	2,042.82
52	2,704	140,608	7.2111	3.7325	.019230769	163.36	2,123.72
53	2,809	148,877	7.2801	3.7563	.018867925	166.50	2,206.18
54	2,916	157,464	7.3485	3.7798	.018518519	169.65	2,290.22
55	3,025	166,375	7.4162	3.8030	.018181818	172.79	2,375.83
56	3,136	175,616	7.4833	3.8259	.017857143	175.93	2,463.01
57	3,249	185,193	7.5498	3.8485	.017543860	179.07	2,551.76
58	3,364	195,112	7.6158	3.8709	.017241379	182.21	2,642.08
59	3,481	205,379	7.6811	3.8930	.016949153	185.35	2,733.97
60	3,600	216,000	7.7460	3.9149	.016666667	188.50	2,827.43
61	3,721	226,981	7.8102	3.9365	.016393443	191.64	2,922.47
62	3,844	238,328	7.8740	3.9579	.016129032	194.78	3,019.07
63	3,969	250,047	7.9373	3.9791	.015873016	197.92	3,117.25
64	4,096	262,144	8.0000	4.0000	.015625000	201.06	3,126.99
65	4,225	274,625	8.0623	4.0207	.015384615	204.20	3,381.31
66	4,356	287,496	8.1240	4.0412	.015151515	207.34	3,421.19
67	4,489	300,763	8.1854	4.0615	.014925373	210.49	3,525.65
68	4,624	314,432	8.2462	4.0817	.014705882	213.63	3,631.68
69	4,761	328,509	8.3066	4.1016	.014492754	216.77	3,739.28
70	4,900	343,000	8.3666	4.1213	.014285714	219.91	3,848.45
71	5,041	357,911	8.4261	4.1408	.014084517	233.05	3,959.19
72	5,184	373,248	8.4853	4.1602	.013888889	226.19	4,071.50
73	5,329	389,017	8.5440	4.1793	.013698630	229.34	4,185.39
74	5,476	405,224	8.6023	4.1983	.013513514	232.48	4,300.84
75	5,625	421,875	8.6603	4.2172	.013333333	235.62	4,417.86
76	5,776	438,976	8.7178	4.2358	.013157895	238.76	4,536.46
77	5,929	456,533	8.7750	4.2543	.012987013	241.90	4,656.63
78	6,084	474,552	8.8318	4.2727	.012820513	245.04	4,778.36
79	6,241	493,039	8.8882	4.2908	.012658228	248.19	4,901.67
80	6,400	512,000	8.9443	4.3089	.012500000	251.33	5,026.55
81	6,561	531,441	9.0000	4.3267	.012345679	254.47	5,153.00
82	6,724	551,368	9.0554	4.3445	.012195122	257.61	5,281.02
83	6,889	571,787	9.1104	4.3621	.012048193	260.75	5,410.61
84	7,056	592,704	9.1652	4.3795	.011904762	263.89	5,541.77
85	7,225	614,125	9.2195	4.3968	.011764706	267.04	5,674.50
86	7,396	636,056	9.2376	4.4140	.011627907	270.18	5,808.80
87	7,569	638,503	9.3274	4.4310	.011494253	273.32	5,944.68
88	7,744	681,472	9.3808	4.4480	.011363636	276.46	6,082.12
89	7,921	704,969	9.4340	4.4647	.011235955	279.60	6,221.14
90	8,100	729,000	9.4868	4.4814	.011111111	282.74	6,361.73
91	8,281	753,571	9.5394	4.4979	.010989011	285.88	6,503.88
92	8,464	778,688	9.5917	4.5144	.010869565	289.03	6,647.61
93	8,649	804,357	9.6437	4.5307	.010752688	292.17	6,792.91
94	8,836	830,584	9.6954	4.5468	.010638298	295.31	6,939.78
95	9,025	857,375	9.7468	4.5629	.010526316	298.45	7,088.22
96	9,216	884,736	9.7980	4.5789	.010416667	301.59	7,283.23
97	9,409	912,673	9.8489	4.5947	.010309278	304.73	7,389.81
98	9,604	941,192	9.8995	4.6104	.010204082	307.88	7,542.96
99	9,801	970,299	9.9499	4.6261	.010101010	311.02	7,697.69
100	10,000	1,000,000	10.0000	4.6416	.010000000	314.16	7,853.98

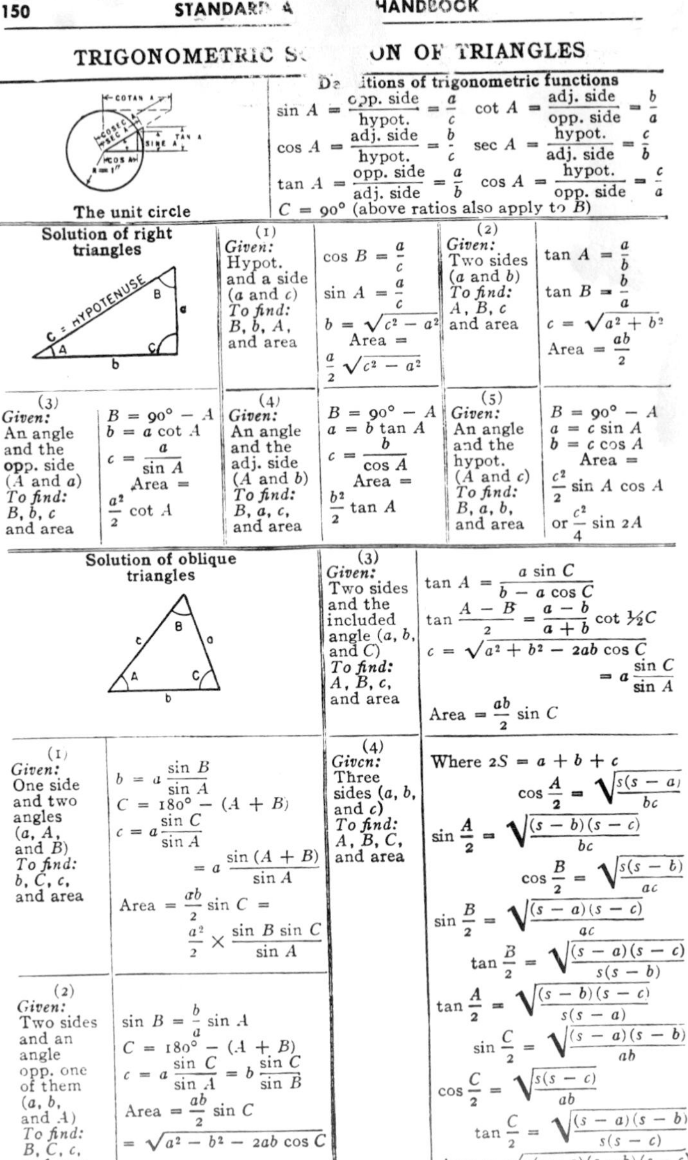

TRIGONOMETRIC S[illegible]ON OF TRIANGLES

D[illegible]itions of trigonometric functions

$\sin A = \dfrac{\text{opp. side}}{\text{hypot.}} = \dfrac{a}{c}$ $\quad \cot A = \dfrac{\text{adj. side}}{\text{opp. side}} = \dfrac{b}{a}$

$\cos A = \dfrac{\text{adj. side}}{\text{hypot.}} = \dfrac{b}{c}$ $\quad \sec A = \dfrac{\text{hypot.}}{\text{adj. side}} = \dfrac{c}{b}$

$\tan A = \dfrac{\text{opp. side}}{\text{adj. side}} = \dfrac{a}{b}$ $\quad \cos A = \dfrac{\text{hypot.}}{\text{opp. side}} = \dfrac{c}{a}$

$C = 90°$ (above ratios also apply to B)

The unit circle

Solution of right triangles

(1) *Given:* Hypot. and a side (a and c). *To find:* B, b, A, and area

$\cos B = \dfrac{a}{c}$

$\sin A = \dfrac{a}{c}$

$b = \sqrt{c^2 - a^2}$

Area $= \dfrac{a}{2}\sqrt{c^2 - a^2}$

(2) *Given:* Two sides (a and b). *To find:* A, B, c and area

$\tan A = \dfrac{a}{b}$

$\tan B = \dfrac{b}{a}$

$c = \sqrt{a^2 + b^2}$

Area $= \dfrac{ab}{2}$

(3) *Given:* An angle and the opp. side (A and a). *To find:* B, b, c and area

$B = 90° - A$

$b = a \cot A$

$c = \dfrac{a}{\sin A}$

Area $= \dfrac{a^2}{2} \cot A$

(4) *Given:* An angle and the adj. side (A and b). *To find:* B, a, c, and area

$B = 90° - A$

$a = b \tan A$

$c = \dfrac{b}{\cos A}$

Area $= \dfrac{b^2}{2} \tan A$

(5) *Given:* An angle and the hypot. (A and c). *To find:* B, a, b, and area

$B = 90° - A$

$a = c \sin A$

$b = c \cos A$

Area $= \dfrac{c^2}{2} \sin A \cos A$ or $\dfrac{c^2}{4} \sin 2A$

Solution of oblique triangles

(1) *Given:* One side and two angles (a, A, and B). *To find:* b, C, c, and area

$b = a \dfrac{\sin B}{\sin A}$

$C = 180° - (A + B)$

$c = a \dfrac{\sin C}{\sin A} = a \dfrac{\sin (A + B)}{\sin A}$

Area $= \dfrac{ab}{2} \sin C = \dfrac{a^2}{2} \times \dfrac{\sin B \sin C}{\sin A}$

(2) *Given:* Two sides and an angle opp. one of them (a, b, and A). *To find:* B, C, c, and area

$\sin B = \dfrac{b}{a} \sin A$

$C = 180° - (A + B)$

$c = a \dfrac{\sin C}{\sin A} = b \dfrac{\sin C}{\sin B}$

Area $= \dfrac{ab}{2} \sin C$

$= \sqrt{a^2 - b^2 - 2ab \cos C}$

(3) *Given:* Two sides and the included angle (a, b, and C). *To find:* A, B, c, and area

$\tan A = \dfrac{a \sin C}{b - a \cos C}$

$\tan \dfrac{A - B}{2} = \dfrac{a - b}{a + b} \cot \tfrac{1}{2}C$

$c = \sqrt{a^2 + b^2 - 2ab \cos C}$

$= a \dfrac{\sin C}{\sin A}$

Area $= \dfrac{ab}{2} \sin C$

(4) *Given:* Three sides (a, b, and c). *To find:* A, B, C, and area

Where $2S = a + b + c$

$\cos \dfrac{A}{2} = \sqrt{\dfrac{s(s - a)}{bc}}$

$\sin \dfrac{A}{2} = \sqrt{\dfrac{(s - b)(s - c)}{bc}}$

$\cos \dfrac{B}{2} = \sqrt{\dfrac{s(s - b)}{ac}}$

$\sin \dfrac{B}{2} = \sqrt{\dfrac{(s - a)(s - c)}{ac}}$

$\tan \dfrac{B}{2} = \sqrt{\dfrac{(s - a)(s - c)}{s(s - b)}}$

$\tan \dfrac{A}{2} = \sqrt{\dfrac{(s - b)(s - c)}{s(s - a)}}$

$\sin \dfrac{C}{2} = \sqrt{\dfrac{(s - a)(s - b)}{ab}}$

$\cos \dfrac{C}{2} = \sqrt{\dfrac{s(s - c)}{ab}}$

$\tan \dfrac{C}{2} = \sqrt{\dfrac{(s - a)(s - b)}{s(s - c)}}$

Area $= \sqrt{s(s - a)(s - b)(s - c)}$

TYPICAL* MECHANICAL PROPERTIES OF WROUGHT ALUMINUM ALLOYS

Alloy and temper	Tension				Hardness	Shear	Fatigue
	Yield strength (set = 0.2 %), psi	Ultimate strength, psi	Elongation, per cent in 2 in. Sheet specimen (1/16 in. thick)	Elongation, per cent in 2 in. Round specimen (1/2 in. dia.)	Brinell, 500-kg. load 10-mm. ball	Shearing strength, psi	Endurance limit, psi
1100-0	5,000	13,000	35	45	23	9,500	5,000
1100-H12	13,000	15,000	12	25	28	10,000	6,000
1100-H14	14,000	17,000	9	20	32	11,000	7,000
1100-H16	17,000	20,000	6	17	38	12,000	8,500
1100-H18	21,000	24,000	5	15	44	13,000	8,500
3003-0	6,000	16,000	30	40	28	11,000	7,000
3003-H12	15,000	18,000	10	20	35	12,000	8,000
3003-H14	18,000	21,000	8	16	40	14,000	9,000
3003-H16	21,000	25,000	5	14	47	15,000	9,500
3003-H18	25,000	29,000	4	10	55	16,000	10,000
2017-0	10,000	26,000	20	22	45	18,000	11,000
2017-T4	40,000	62,000	20	22	100	36,000	15,000
Alclad 2017-T4	33,000	56,000	18	..	...	32,000	
2117-T4	24,000	43,000	..	27	70	26,000	13,500
2024-0	10,000	26,000	20	22	42	18,000	12,000
2024-T4	45,000	68,000	19	22	105	41,000	18,000
2024-T36	55,000	70,000	13	..	116	42,000	
Alclad 2024-T	41,000	62,000	18	..	...	40,000	
Alclad 2024-T36	50,000	66,000	11	..	...	41,000	
5052-0	14,000	29,000	25	30	45	18,000	17,000
5052-H12	26,000	34,000	12	18	62	20,000	18,000
5052-H14	29,000	37,000	10	14	67	21,000	19,000
5052-H16	34,000	39,000	8	10	74	23,000	20,000
5052-H18	36,000	41,000	7	8	85	24,000	20,500
6053-0	7,000	16,000	25	35	26	11,000	7,500
6053-T4	20,000	33,000	22	30	65	20,000	10,000
6053-T6	33.000	39,000	14	20	80	24,000	11,000
6061-0	8,000	18,000	22	..	30	12,500	8,000
6061-T4	21,000	35,000	22	..	65	24,000	12,500
6061-T6	39,000	45,000	12	..	95	30,000	12,500
7075-0	15,000	33,000	17	16	60	22,000	
7075-T6	72,000	82,000	11	11	150	49,000	
Alclad 7075-0	14,000	32,000	17	—	—	22,000	
7075-T6	67,000	76,000	11	—	—	46,000	

* These values are *not* guaranteed.

CONDITIONS FOR HE[illegible] [illegible]REATMENT OF ALUMIN[illegible] AL[illegible]YS

Alloy	Solution heat-treatment			Precipitation heat-treat[illegible]		
	Temp., °F	Quench	Temper desig.	Temp., °F	Time of aging	Te[illegible] d[illegible]
2017	930–950	Cold water	T4			T
2117	930–950	Cold water	T4			[illegible]
2024	910–930	Cold water	T4			T
6053	960–980	Water	T4	445-455	1-2 hr	T5
				or		
				345–355	8 hr	T6
6061	960–980	Water	T4	315–325	18 hr	T6
				or		
				345–355	8 hr	T6
7075	870	Water		250	24 hr	T6

NOMINAL COMPOSITION OF WROUGHT ALUMINUM ALLOYS*

Alloy	Per cent of alloying elements—aluminum and normal impurities constitute remainder								
	Copper	Silicon	Manganese	Magnesium	Zinc	Nickel	Chromium	Lead	Bismuth
1100									
3003	....	....	1.2						
2011	5.5	....	...	...	....	...	....	0.5	0.5
2014	4.4	0.8	0.8	0.4					
2017	4.0	....	0.5	0.5					
2117	2.5	....	...	0.3					
2018	4.0	....	...	0.5	....	2.0			
2024	4.5	....	0.6	1.5					
2025	4.5	0.8	0.8						
4032	0.9	12.5	....	1.0	....	0.9			
6151	....	1.0	...	0.6	....	...	0.25		
5052	....	....	...	2.5	....	...	0.25		
6053	....	0.7	...	1.3	....	...	0.25		
6061	0.25	0.6	...	1.0	....	...	0.25		
7075	1.6	....		2.5	5.6		0.3		

TYPICAL SOAKING TIMES FOR HEAT-TREATMENT

Thickness, in.	Time, minutes,
Up to .032	30
.032 to ⅛	30
⅛ to ¼	40
Over ¼	60

NOTES: Soaking time starts when the metal (or the molten bath) reaches a temperature within the range specified above.

APPROXIMATE TEMPERATURES FOR HEAT-TREATMENT OF STEELS

	Annealing temp., °F	Hardening temp., °F	Quench	Tensile strength, psi	Tempering temp., °F	Tensile strength, psi
S.A.E. 1025	1575–1650	1575–1650	Water or brine	55,000	Does not harden sufficiently to require tempering	
S.A.E. 1035	1575–1675	1525–1575	Water-oil—if less than ⅜ thick	In the hardened condition brittleness and strains are present and are undesirable for structural parts	400–800	100,000
S.A.E. 1045	1475–1650	1475–1525	Water-oil—if less than ⅜ thick		400–950	110,000
S.A.E. 1095	1475–1550	1400–1450 1500–1550	Water Oil		450–800	220,000
S.A.E 2330	1500–1600	1430–1500	Oil		1050 950 800	100,000 125,000 150,000
S.A.E. X-4130 / X-4135	1600–1700 Cool furnace to 1100 before removing. Tensile strength, 90,000	1550–1650	Oil		1075 1000 950 650	125,000 150,000 180,000 200,000
S.A.E. 4140	1600–1700	1525–1625	Oil		1250 1100 650	100,000 125,000 200,000
S.A.E. X-4340	1500–1600	1475–1525	Oil		1200 1050 950 850	125,000 150,000 180,000 200,000
S.A.E. 3140	1550–1650	1475–1525	Oil		1050 950 800	100,000 150,000 180,000
S.A.E. 3250	1500–1600	1425–1475	Oil		1100 950 800	150,000 180,000 200,000
S.A.E. 3435	1550–1650	1400–1450	Oil		1000 900 800	150,000 180,000 200,000

COLOR CHART FOR STEEL AT VARIOUS TEMPERATURES

Color	Metal Temp., °F
Faint red	900
Blood red	1050
Dark cherry	1075
Medium cherry	1250
Cherry or full red	1375
Bright red	1550
Salmon	1650
Orange	1725
Lemon	1825
Light yellow	1975
White	2200
Dazzling white	2350

COLOR CHART FOR VARIOUS TEMPERING TEMPERATURES OF CARBON STEELS

Oxide Color	Metal Temp., °F
Pale yellow	428
Straw	446
Golden yellow	469
Brown	491
Brown, dappled with purple	509
Purple	531
Dark blue	550
Bright blue	567
Pale blue	610

COLORS, FLUID LINES IDENTIFICATION

All bands shall be 1 in. wide and shall encircle the tube.

Bands shall be located near each end of the tube and at such intermediate points as may be necessary to follow through the system.

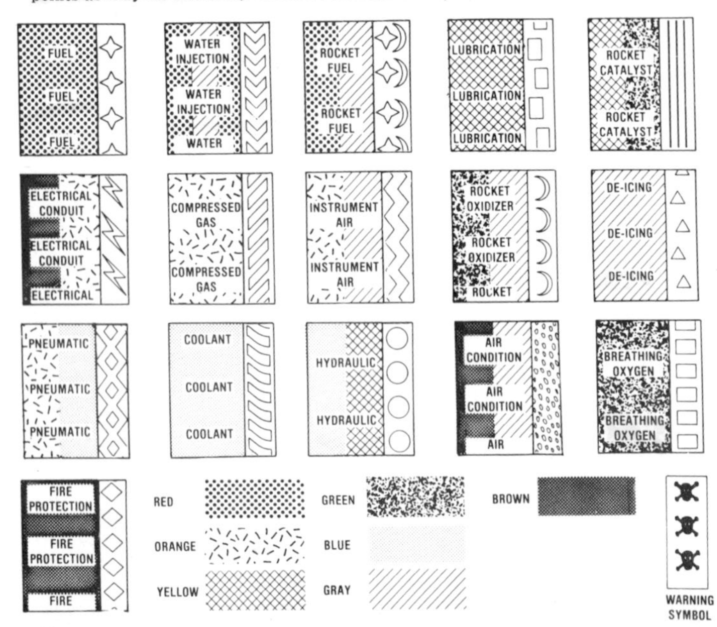

APPROXIMATE RELATIONS BETWEEN HARDNESS AND TENSILE STRENGTH OF S.A.E. STEELS*

Brinell		Rockwell			
Dia. in mm. 3,000-kg load, 10-mm ball	Hardness number	C scale, 150-kg load, 120° diamond cone	B scale, 100-kg load, 1/16-in. dia. ball	Shore scleroscope number	Tensile strength, 1,000 psi
2.20	780	70	. . .	106	384
2.25	745	68	. . .	100	368
2.30	712	66	. . .	95	352
2.35	682	64	. . .	91	337
2.40	653	62	. . .	87	324
2.45	627	60	. . .	84	311
2.50	601	58	. . .	81	298
2.55	578	57	. . .	78	287
2.60	555	55	. . .	75	276
2.65	534	53	. . .	72	266
2.70	514	52	. . .	70	256
2.75	495	50	. . .	67	247
2.80	477	49	. . .	65	238
2.85	461	47	. . .	63	229
2.90	444	46	. . .	61	220
2.95	429	45	. . .	59	212
3.00	415	44	. . .	57	204
3.05	401	42	. . .	55	196
3.10	388	41	. . .	54	189
3.15	375	40	. . .	52	182
3.20	363	38	. . .	51	176
3.25	352	37	. . .	49	170
3.30	341	36	. . .	48	165
3.35	331	35	. . .	46	160
3.40	321	34	. . .	45	155
3.45	311	33	. . .	44	150
3.50	302	32	. . .	43	146
3.55	293	31	. . .	42	142
3.60	285	30	. . .	40	138
3.65	277	29	. . .	39	134
3.70	269	28	. . .	38	131
3.75	262	26	. . .	37	128
3.80	255	25	. . .	37	125
3.85	248	24	. . .	36	122
3.90	241	23	100	35	119
3.95	235	22	99	34	116
4.00	229	21	98	33	113
4.05	223	20	97	32	110
4.10	217	. .	96	31	107
4.15	212	. .	96	31	104
4.20	207	. .	95	30	101
4.25	202	. .	94	30	99
4.30	197	. .	93	29	97
4.35	192	. .	92	28	95
4.40	187	. .	91	28	93
4.45	183	. .	90	27	91
4.50	179	. .	89	27	89
4.55	174	. .	88	26	87
4.60	170	. .	87	26	85
4.65	166	. .	86	25	83
4.70	163	. .	85	25	82
4.75	159	. .	84	24	80
4.80	156	. .	83	24	78
4.85	153	. .	82	23	76

* Emphasis is laid on the fact that this table gives an approximate relationship of Brinell, Rockwell, and scleroscope values. It is impossible to give more than an approximate relationship owing to the inevitable influence of size, mass, composition, and method of heat-treatment. Where more precise factors are required, they should be especially developed for each steel composition, heat-treatment, and part. (This table was reproduced by permission of the Society of Automotive Engineers.)

(*Continued*)

APPROXIMATE RELATIONS BETWEEN HARDNESS AND TENSILE STRENGTH OF S.A.E. STEELS.†

(Continued)

Brinell		Rockwell		Shore scleroscope number	Tensile strength, 1,000 psi
Dia. in mm. 3,000-kg load, 10-mm ball	Hardness number	C scale, 150-kg load, 120° diamond cone	B scale, 100-kg load, 1/16-in. dia. ball		
4.90	149	..	81	23	75
4.95	146	..	80	22	74
5.00	143	..	79	22	72
5.05	140	..	78	21	71
5.10	137	..	77	21	70
5.15	134	..	76	21	68
5.20	131	..	74	20	66
5.25	128	..	73	20	65
5.30	126	..	72	...	64
5.35	124	..	71	...	63
5.40	121	..	70	...	62
5.45	118	..	69	...	61
5.50	116	..	68	...	60
5.55	114	..	67	...	59
5.60	112	..	66	...	58
5.65	109	..	65	...	56
5.70	107	..	64	...	55
5.75	105	..	62	...	54
5.80	103	..	61	...	53
5.85	101	..	60	...	52
5.90	99	..	59	...	51
5.95	97	..	57	...	50
6.00	95	..	56	...	49

TUBING, ALUMINUM ALLOY (52S)—STANDARD SIZES FOR FLUID LINES

O.D., in.	Wall thickness, in.	Wall thickness (optional), in.	O.D., in.	Wall thickness, in.
1/8	.032	....	1	.049
3/16	.032	....	1 1/4	.058
1/4	.032	....	1 1/2	.058
5/16	.032	.042	1 3/4	.058
3/8	.032	.049	2	.065
1/2	.042	.058	2 1/2	.065
5/8	.042	....	3	.065
3/4	.049			

TUBING, STEEL, CORROSION RESISTING—STANDARD SIZES FOR FLUID LINES

Material: Spec. AN–WW–T–858.

O.D., in.	Wall thickness, in.	O.D., in.	Wall thickness, in.
1/8	.022	3/4	.035
3/16	.022	1	.035 (.049 optional)
1/4	.022	1 1/4	.049
5/16	.028	1 1/2	.049
3/8	.028	1 3/4	.049
1/2	.028	2	.049
5/8	.035		

...IT, RIGID, ALUMINUM—STANDARD SIZES FOR

O.D., in.	Wall thickness, in.	O.D., in.	Wall thickness, in.	O.D., in.	Wall thickness, in.
[illegible]	.022	5/8	.028	1 1/2	.042
[illegible]	.022	3/4	.032	1 3/4	.042
[illegible]	.022	1	.032	2	.049
[illegible]	.028	1 1/4	.042	2 1/2	.058

Material: Federal Spec. WW–T–783, 1/2H.

TUBING, STEEL, CHROME-MOLYBDENUM SEAMLESS, ROUND—STANDARD SIZES FOR

O.D., in.	Wall thickness											
	.022	.028	.035	.049	.058	.065	.083	.095	.120	.134	.156	.188
3/16	..	..	x									
1/4	x	x	x	x	x	x						
5/16	..	x	x	x	x	x	..	x				
3/8	..	x	x	x	x	x	x	x				
7/16	..	x	x	x	x	x	x	x				
1/2	..	x	x	x	x	x	x	x				
9/16	..	..	x	x	..	x	..	x				
5/8	..	x	x	x	x	x	x	x	x			
3/4	..	x	x	x	x	x	x	x	x			
7/8	..	x	x	x	x	x	..	x	x			
1	..	..	x	x	x	x	x	x	x	..	..	x
1 1/8	..	..	x	x	x	x	x	x	x			
1 1/4	..	..	x	x	x	x	x	x	x			
1 3/8	..	..	x	x	x	x	x	x	x			
1 1/2	..	..	x	x	x	x	x	x	x	..	..	x
1 5/8	..	..	..	x	x	x	x	x	x			
1 3/4	..	..	x	x	x	x	x	x				
1 7/8	..	..	..	x	x	x	x	x	x			
2	..	..	x	x	x	x	x	x	x	..	..	x
2 1/4	..	..	..	..	x	x	x	x	x	..	..	x
2 1/2	..	..	..	..	..	x	x	x	x			
2 3/4	..	..	..	..	..	x	x	..	x			
3	..	..	..	..	..	..	x					
3 1/4	..	..	..	..	..	..	x					
3 1/2	..	..	..	..	..	..	..	x				
3 3/4	..	..	..	..	..	..	..	x				
4	..	..	..	..	..	..	..	..	x			
4 1/4	..	..	..	..	..	..	..	..	..	x		
4 1/2	..	..	..	..	..	..	..	..	..	..	x	
4 3/4	..	..	..	..	..	..	..	..	..	..	..	x

INDEX

DECIMAL EQUIVALENTS

OF PARTS OF AN INCH

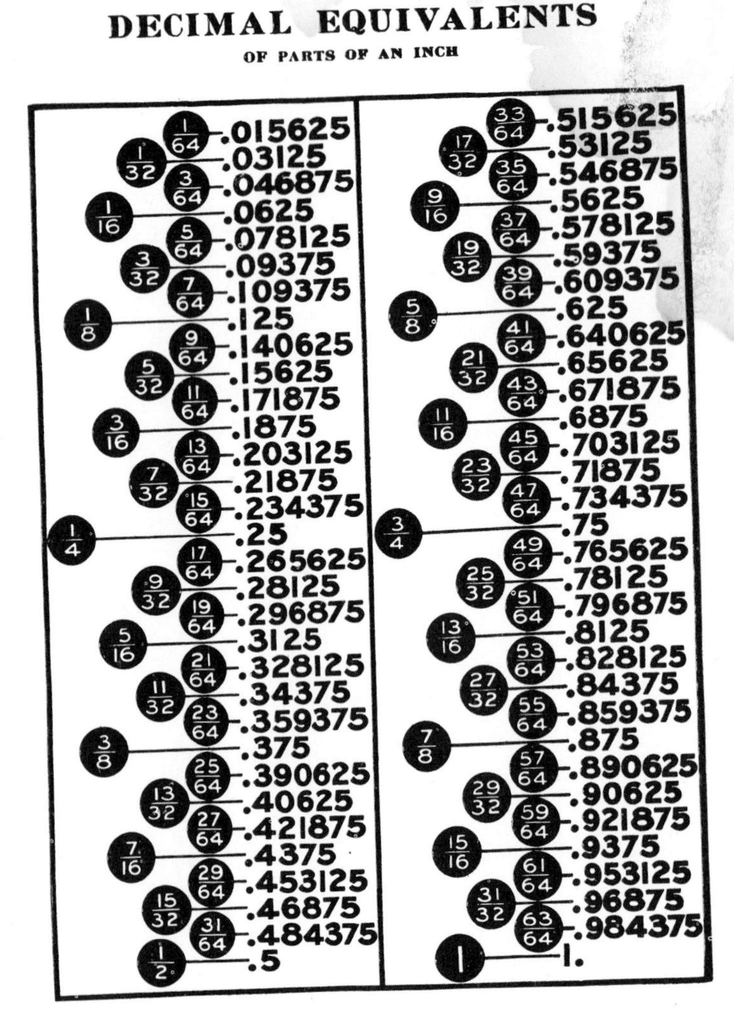

Fraction	Decimal
1/64	.015625
1/32	.03125
3/64	.046875
1/16	.0625
5/64	.078125
3/32	.09375
7/64	.109375
1/8	.125
9/64	.140625
5/32	.15625
11/64	.171875
3/16	.1875
13/64	.203125
7/32	.21875
15/64	.234375
1/4	.25
17/64	.265625
9/32	.28125
19/64	.296875
5/16	.3125
21/64	.328125
11/32	.34375
23/64	.359375
3/8	.375
25/64	.390625
13/32	.40625
27/64	.421875
7/16	.4375
29/64	.453125
15/32	.46875
31/64	.484375
1/2	.5
33/64	.515625
17/32	.53125
35/64	.546875
9/16	.5625
37/64	.578125
19/32	.59375
39/64	.609375
5/8	.625
41/64	.640625
21/32	.65625
43/64	.671875
11/16	.6875
45/64	.703125
23/32	.71875
47/64	.734375
3/4	.75
49/64	.765625
25/32	.78125
51/64	.796875
13/16	.8125
53/64	.828125
27/32	.84375
55/64	.859375
7/8	.875
57/64	.890625
29/32	.90625
59/64	.921875
15/16	.9375
61/64	.953125
31/32	.96875
63/64	.984375
1	1.